ELEMENTARY CLIMATE PHYSICS

# Elementary Climate Physics

F.W. TAYLOR

*Department of Physics,*
*University of Oxford, UK*

OXFORD
UNIVERSITY PRESS

*This book has been printed digitally and produced in a standard specification*
*in order to ensure its continuing availability*

# OXFORD
UNIVERSITY PRESS

Great Clarendon Street, Oxford OX2 6DP

Oxford University Press is a department of the University of Oxford.
It furthers the University's objective of excellence in research, scholarship,
and education by publishing worldwide in

Oxford New York

Auckland Cape Town Dar es Salaam Hong Kong Karachi
Kuala Lumpur Madrid Melbourne Mexico City Nairobi
New Delhi Shanghai Taipei Toronto
With offices in
Argentina Austria Brazil Chile Czech Republic France Greece
Guatemala Hungary Italy Japan South Korea Poland Portugal
Singapore Switzerland Thailand Turkey Ukraine Vietnam

Oxford is a registered trade mark of Oxford University Press
in the UK and in certain other countries

Published in the United States
by Oxford University Press Inc., New York

ISBN 978-0-19-856733-2

Printed and bound in Great Britain by CPI Antony Rowe, Chippenham and Eastbourne

# Preface

Understanding the Earth's climate and predicting its future behaviour is, first and foremost, a problem in Physics. Climate Physics is a new subject on many university syllabuses, but one that has strong echoes of the traditional precursor of modern Physics, Experimental Philosophy, with its emphasis on understanding the natural world around us. During the twentieth century, much of the emphasis had swung towards the physics of the very small, with the advent of quantum theory and elementary particle physics, and the very large, with the growth of extra-galactic astronomy and cosmology. Today, the pendulum is swinging back, and physics on the scale of direct human experience is reclaiming its place as a topic of compelling interest, in schools and universities, and in everyday life as well.

The reason for the renaissance is clear: at last we are in a position to begin to answer some of the most fundamental questions facing humankind. How did our Earth develop, and come to be the way it is? How is its observed behaviour to be explained in terms of scientific theory, and in terms of the parallel behaviour of the neighbouring planets of the Solar System and beyond? How did life evolve? What makes a planet support life, and is habitability sustainable in the face of climate change? How accurately and at what range can we forecast weather and climate? And so forth. Even at a very basic level, the key to the answer of questions such as these lies in modern studies of the physical world, of which the climate system is now seen to be a key part.

Climate Physics is a modern subject in that the relevant data bases on the physical states of the atmosphere and ocean, their planetary-scale history and evolution, the global measurement systems, and the sophisticated computer models, which collectively make quantitative studies and predictions possible, are all recent innovations. At the same time, interest in understanding the climate has received an enormous boost from the concern generated by the realization that rapid climate change, much of it forced by the relentless increase in population and industrialization, is potentially a serious threat to the quality of life on Earth. Our ability to resist and overcome any such threat depends directly on our ability to understand what physical effects are involved and to predict how trends may develop. In an introductory course, we want to clarify the basics, topic by topic, and see how far we can get by applying relatively simple Physics to the climate problem. This provides a foundation for more advanced work, which we can identify and appreciate at this level although of course a full treatment requires more advanced books, of which there are many.

In January 2004 a new unit entitled *Climate Physics* was introduced into the mandatory part of the undergraduate Physics syllabus at Oxford University. The topics covered, under the banner of *Physics of Atmosphere and Ocean,* had previously been a final-year option taken by students who had some expectation of continuing in the field, either as a graduate student or in some other capacity, or at least some special interest in the subject. Now we had to cater for younger, less-experienced students coming to the subject for the first time, and leaving it a few weeks later, most of them for ever.

In preparing the lectures for this course, it rapidly became clear that, despite a burgeoning interest worldwide in the basic science of climate and a corresponding increase in the literature, there was no suitable textbook in existence for students at this relatively low level. A number of excellent books deal with the physics of the atmosphere and/or ocean, but these are invariably aimed at graduate students and researchers, or possibly final-year undergraduates specializing in the subject. A student coming to the subject as part of a general Physics course needs a very basic introduction to what is important in the climate field and how it relates to the rest of Physics that he or she is learning at the same time. The treatment has to be elementary, quantitative, and empirical so that it can be applied to simple numerical problems in tutorial classes and examinations.

I have also found that the majority of books in climate-related fields, at least those that address the physics of climate problems, are oriented towards geophysical fluid dynamics. This is a complex and specialized topic, which forms a separate module at Oxford, with its own textbooks. The few books that do deal with thermodynamics, radiative energy balance, modelling, and measurements systems, and that are not too advanced, tend to be written for interdisciplinary students and do not relate well to the fundamental physics.

This new book was therefore conceived and aimed squarely at undergraduates taking physical sciences courses who require an elementary treatment, but with the expectation that it could also provide a useful introduction to the subject for first-year graduate students and others entering the field for the first time. The style is deliberately didactical, with a certain amount of repetition of key definitions and concepts between chapters both for emphasis and to aid the student who uses the text selectively or for revision. References to some appropriate further reading have been provided that will allow the student who wants to delve further into the physics or applications of any of the areas treated in a basic way here. Further reading on all of the topics covered in this introduction will be found in the chapters that follow. At the end of each chapter is a guide to more advanced books on each subject, and a set of questions. The latter are provided as samples of what might be found in examinations at the undergraduate, non-specialist level, and can also be used by the student to test his or her comprehension of the material if they are attempted after the chapter has been studied, but without referring back to the text until the answer is required.

Thanks are expressed to my Oxford colleagues who have taught similar topics in the past and have been generous with their material, especially Jim Williamson, Clive Rodgers, and David Andrews. They also provided many useful comments on the manuscript, and so significantly improved it. The figures were specially drawn by Dr D.J. Taylor, to whom I am extremely grateful for excellent work. The attribution of ideas and data contained in the text and figures is often difficult in a book as basic as this; precedence is difficult to establish and in many cases an attempt to make an attribution would be confusing. I have sought therefore to limit references to cases where information useful to the student is conveyed, and hope a blanket acknowledgement will be accepted by the many researchers in universities, institutes, government departments, and space agencies, worldwide, past and present, who contributed to the material covered by this tutorial introduction to the field. Finally, I thank the Senior Physical Sciences Editor of Oxford University Press, Sonke Adlung, who has been most helpful and supportive.

F.W. Taylor
*Atmospheric, Oceanic and Planetary Physics,*
*Clarendon Laboratory, Oxford.*
*January 2005*

# Contents

Contents

Contents

# Contents

# The climate system

1

## 1.1 Introduction: A definition of climate

The climate 'system' consists primarily of the land, ocean, and ice on the surface of the Earth; the atmosphere that overlies it; and the radiation from the Sun that provides energy. All of these interact, to produce the conditions on and around the surface of our planet that we call the *climate*. While the term can refer to the mean state of the whole system, averaged in space and time, the emphasis is usually on the parameters that most affect life on the planet, especially the surface temperature, and its global and seasonal variation. Just as important as temperature, although probably more stable at the present time, are the oxygen abundance in the atmosphere, and the mean surface pressure. Factors such as cloud cover, wind field, and precipitation are subject to natural fluctuations and long-term change, and are generally considered part of the climate.

Climate Physicists seek to be able to calculate the state of the climate from an understanding of the physical laws that govern it, given some basic boundary conditions. We need to understand how stable it is, and how it may change in response to external or internal forcing. Examples of external forcing would be solar variability, or the impact of a large meteor or comet on the Earth, and an example of internal forcing would be changes in the minor constituent abundances in the atmosphere, due to volcanism, or to pollution by industrial and other human activities.

To understand and predict the climate, we must draw from a wide range of disciplines in Physics. The following list, although far from exhaustive, contains the most obvious of these:

(1) The processes in the Sun that produce its electromagnetic spectrum;
(2) the interactions between solar photons and atmospheric molecules;
(3) the effect of solar radiation on atmospheric composition (photochemistry);
(4) the thermodynamics of the atmosphere;
(5) the fluid dynamics of the atmosphere and ocean;
(6) radiative transfer in the atmosphere;
(7) cloud physics;
(8) geophysical measurements, including remote sensing observations from satellites;
(9) numerical climate modelling, including predictive coupled models of ocean and atmosphere.

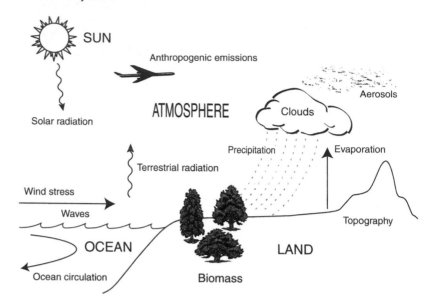

**Fig. 1.1**
Schematic of some of the principal components and relationships in the climate system.

[1] Lewis Fry Richardson (1881–1953) was the father of modern meteorology, that is, he pioneered the use of physical equations and calculations to forecast the weather, although the necessary computing power was not available in 1922—he had to use a room full of people with slide rules. He estimated that he would need at least 60,000 people to calculate the weather before it actually arrived.

The basic physics involved in each of these will all be covered in some detail in the following chapters. Figure 1.1 illustrates the climate system and some of the more important interactions that take place.

First, let us return to what we mean by climate. The adage, 'climate is what you expect, weather is what you get'—attributed in climate circles to the British meteorologist L.F. Richardson,[1] although he may have been quoting Mark Twain—usefully conveys the basic differences between the two. We represent the climate as the mean state of the geosystem, especially at the surface, after diurnal (day–night), seasonal (winter–summer) interannual (year-to-year) and random fluctuations have been averaged out. Thus weather systems, which have lifetimes of only a few days, contribute to the climate only in a statistical mean sense, and when we speak of climate *change*, in this definition we mean something that cannot be detected except by measurements that cover several years at least.

Temperature is the key climate variable, and when we choose a single parameter to represent climate and climate change, it is usually the average temperature of the Earth's surface. Often the distribution of temperature with height and across the globe is of interest, and how this varies on seasonal time scales and from year to year or decade to decade. We are also concerned with quantities such as humidity, precipitation, cloud amount, wind force, and direction, not as discrete weather phenomena, but in terms of their statistical behaviour over time. For instance, we wish to know how frequently extreme events—hurricanes, for example, or droughts—occur each year, and how that may change in the longer term, whether or not mean surface temperature is also changing.

Weather forecasting, on timescales of a few days, is complex and demands vast resources. At present even short-term forecasts are sometimes inaccurate, although a steady improvement is being forged with better models, better data, and faster computers. Despite this, forecasting the weather years, decades or longer into the future remains impossible, and so far as we know it always will be, if we demand the same sort of detail we currently find in two- or three-day forecasts. If, however, we reduce the requirements for spatial and temporal resolution to say continental and annual (or at least seasonal) scales respectively, then we believe we have a chance of predicting the climate, although the techniques are still in their infancy, as we shall see.[2] The first step is to be able to build models that show how the present climate arises (nowcasting) and to use them successfully to account for past variations (hindcasting), before facing the daunting task of predicting the future.

These are formidable tasks, which we tackle in a variety of ways, including seeking a better understanding of the processes involved, new and improved measurements, and ever more sophisticated computer models. First we want to understand the basic physics, and to learn how it all fits together to give a basic picture of the climate and what factors are important in altering or stabilizing it. The thing that makes the climate a 'system' is not just its complexity, or the diversity of factors that contribute: it is also the interactions that take place, such as those in the atmosphere among radiation, chemistry, and dynamics. For example, an increase in the surface temperature will increase evaporation and therefore the atmospheric humidity, probably increasing the cloud amount in turn, with a consequent effect on the solar heating at the surface and therefore its temperature. The dependencies in loops like this are often mutual, introducing both positive and negative feedback into climate states and cycles, and making it difficult to find the balance point.

Assuming that the equilibrium state of the climate can be calculated for a given set of boundary conditions still leaves questions about the stability and uniqueness of this state. Complex systems with feedback loops tend to exhibit chaotic behaviour and the climate certainly shows plenty of evidence of this. Large shifts in the past, some of which seem to have taken place very quickly, and then reversed, suggest that the climate may also have multiple quasi-stable equilibrium states.[3] To model its behaviour we need to understand and parameterize the physics[4] involved, but very comprehensive measurements are essential before the various processes and interactions at work can be distinguished from each other and their individual roles appreciated under all conditions. Very detailed numerical models are required to represent them accurately, and even then it can be difficult just to separate chaotic behaviour from that which is complex and may look chaotic, but which nevertheless has organized elements. Problems of *attribution* (deciding how an observed phenomenon arises), *simulation* (representing phenomena by mathematical formulations of physical laws), and *predictability* (modelling time-dependent behaviour into the future) are at the heart of modern climate research.

[2] A mundane example illustrates the difference between weather and climate forecasting. We cannot predict reliably whether it will rain at a football match in a given town seven days hence, but we can be reasonably certain it will rain somewhere in the country during the following month. In other words, by demanding less detail, confidence can be increased. A key question in climate forecasting is how far we can go, in farsightedness and precision, in theory as well as in practice.

[3] 'Chaotic' behaviour means that the state of the system is not predictable a finite time ahead, no matter how well the input parameters are determined. A stable state is one in which there is negative feedback in response to any change, tending to restore the original conditions; 'quasi-stable' means that a realistic perturbation can overcome the restoring force and produce large changes. The current climate seems to be in this category, since it has remained the same, with small chaotic fluctuations, for thousands of years, although we know it has also undergone large changes in the more distant past (the Ice Ages).

[4] 'Parameterizing the physics' means summarizing and representing complicated processes by a small number of variables (parameters), so that they become tractable in numerical models. The use of single quantity, the albedo, to represent the reflectivity of the Earth, with its complex variations (see Section 1.10), is an example of this.

## 1.2 Solar radiation and the energy budget of the Earth

The source of virtually all the energy in the climate system is radiation from the Sun (see Chapter 2). We must therefore consider how the Sun operates, how the radiation it produces varies with wavelength (i.e. the solar *spectrum*), the amount of energy that reaches an imaginary sphere at the mean distance of the Earth from the Sun (equal to $1.37\,kW\,m^{-2}$, called the *solar constant*), and whether it really is constant in the current epoch.

The Sun, like other stars, experiences evolution on very long time scales. More relevant to our present focus on climate, it also goes through shorter-term cycles, for example, that marked by varying sunspot activity, during which its output of energy changes slightly but perceptibly. Since measurements began, the observed variation in solar output is only a fraction of 1% (see Chapter 2 and Section 10.3). A much larger effect results from the fact that the distance of the Earth from the Sun varies during the year, because the orbit is not quite circular but has an eccentricity of about 3%. Currently, this causes the flux at the Earth to range from $1{,}435\,W\,m^{-2}$ on 3 January to about $1{,}345\,W^{-2}$ on 3 July.[5]

While the effect of this 6% fluctuation in the insolation[6] due to the orbital eccentricity can be detected with instruments, it escapes notice in everyday life because, at any particular location on the planet, the climate is dominated by the normal seasonal cycle produced by the tilt of the Earth's axis of rotation. This quantity, called the obliquity, is about 23°, so the angle at which sunlight illuminates the surface at a latitude of, say, 45°, varies from 22° to 68° from midsummer to midwinter. Since the energy per unit area at the surface depends on the cosine of the incidence angle, the obliquity is responsible for a seasonal variation in mean daily insolation of more than 100% at this latitude. It is also responsible for the phenomenon of the 'polar night', in which the sunfall is zero for 24 h or more in the middle of the winter poleward of 67° latitude (the Arctic and Antarctic circles).

An important, but more elusive, climate variable is the fraction of the incident solar energy that is actually *absorbed* by the planet as a whole (Fig. 1.2). The fraction that is reflected or scattered back into space without being absorbed, which in the case of the Earth is about 0.30, is called the *albedo* of the planet.[7] The fraction $(1 - albedo)$ of the solar constant that is available to power the climate system of the Earth can vary by a large factor because the albedo is largely determined by the amount and type of clouds present at the time. This of course fluctuates considerably and it is difficult to measure global cloud properties on a sufficiently broad and systematic basis.

Nor is the Earth's albedo an easy quantity to measure directly, as it requires not only a platform in space but measurements over all emission

[5] Note that this is not a variation in the solar constant, which is a function only of the output from the Sun, since it is defined in terms of the *mean* Earth–Sun distance.

[6] 'Insolation' (sometimes called 'sunfall') is a general term that refers to the actual energy per unit area reaching a specific location, usually on the Earth's surface, in a given situation. Unlike the solar constant, which is in fact *almost* constant (see Section 2.5), the surface insolation varies substantially with latitude, time of day, season, and cloud cover, and goes to zero at night. The insolation at the top of the atmosphere equals the solar constant when the Earth is at its mean distance from the Sun.

[7] A perfect reflector has an albedo of unity and a perfect absorber has an albedo of zero. Albedo refers to the fraction of the *total* solar energy, that is, at all wavelengths and in all directions within a hemisphere, which is returned to space without being absorbed. Strictly, this is called the Bond albedo (after the American astronomer George Bond, who introduced the concept in 1860). When we want to talk about the fraction of energy returned at a specific wavelength and/or in a specific direction from a planet, a cloud, or some other body, we use the usual term reflectivity, as discussed in Section 6.4.

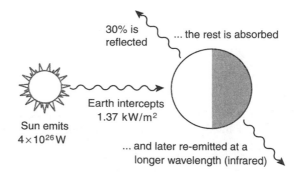

**Fig. 1.2**
The Earth's climate is driven by energy from the Sun. The biggest uncertainty is the planet's albedo, the fraction of the incoming energy that is reflected back into space.

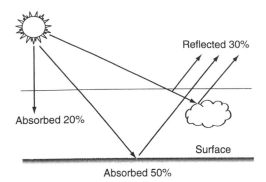

**Fig. 1.3**
Of the energy from the Sun that the Earth intercepts, roughly 30% is reflected (mainly by clouds, but also by reflection from the surface and scattering in the atmosphere), 20% is absorbed in the atmosphere (mainly by minor constituents), and 50% reaches the surface. Although reflection at a single angle is shown for simplicity, backscattering in all directions contributes to the total of 30%.

angles, all wavelengths, and over the whole of the illuminated surface of the planet, simultaneously. Obviously, this is not possible with a single satellite, or any finite number of satellites, and, with current measurement systems, even a reasonably comprehensive estimate is difficult.[8] We shall see in Chapter 9, however, that there is some evidence that the monthly mean value varies by as much as 20% during the year, with long-term trends that are almost completely unknown at present. Such a large variability, if confirmed, would dominate all other uncertainties in the energy balance budget of the Earth, the fundamental equation on which the climate ultimately depends.

On average, about half of the energy from the Sun that arrives at the Earth reaches the surface (Fig. 1.3). The remainder is either reflected back into space, mainly by clouds, but also by reflection from the surface and scattering in the atmosphere, or is absorbed in the atmosphere. Most of the atmospheric absorption is due to *minor* constituents, that is, those present in very small proportions, since the major constituents, nitrogen and oxygen, are largely transparent to sunlight except at very short (ultraviolet, UV) wavelengths. The reasons why some gases interact with visible and infrared radiation while others do not are a consequence of the symmetry properties of the molecule, and will be considered in Chapter 7.

Since the Earth as a whole continuously absorbs $\sim$70% of the energy from the Sun, amounting to a power input of about $2 \times 10^{17}$ W, energy must be lost at the same rate in order to maintain an overall equilibrium. This is accomplished also by radiation, this case at long infrared wavelengths.[9] All of the absorbed energy must ultimately be returned to space, since we observe that the planet neither heats up nor cools down on average over long periods of time. This is true even if the phenomenon of 'global warming' is taking place, since this refers to warming of the surface due to a change in the temperature structure of the atmosphere, as discussed in Chapter 6, and not to an increase in the rate of heating or cooling of the Earth as a whole.

The mean energy budget must balance, not only for the entire planet but also for its various regions, from the largest to the smallest scale. In particular, since the equatorial regions receive much more energy from the

[8] A satellite can measure the albedo at one place, and for one solar incidence angle and one direction of reflection. To cover the whole planet in all spatial directions on a time scale faster than the rate at which cloud cover is changing would need many satellites, in many different orbital planes. They would also need to orbit a long way from the planet, so each one could cover a large fraction of the globe within its field-of-view. Given the importance of knowing the albedo accurately for climate studies, it is not surprising that such a network of satellites has been studied, and it may not be long before a version is implemented.

[9] Note the distinction between the $\sim$30% of solar energy that is returned to space by reflection and scattering without being absorbed, and normally without any change of wavelength, and the remainder, which is absorbed and produces heating of the surface and atmosphere. These go on to emit completely unrelated photons at longer wavelengths to achieve energy balance in the system as a whole.

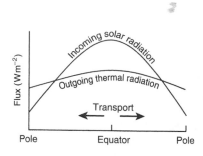

**Fig. 1.4**
The solar flux (radiant energy per unit area) falling on the Earth is greatest at the equator and falls away towards the poles, because of the Earth's spheroidal shape. The flux of thermal radiation back into space depends only on the emitting temperature, and has a smaller dependence on latitude. The difference between the 'energy in' and 'energy out' curves is balanced by massive transfer of heat from low to high latitudes, primarily by transport in the ocean and the atmosphere.

Sun than the polar regions, while the cooling from the terrestrial globe is more nearly the same everywhere, the ocean and atmosphere must circulate in such a way as to transport heat polewards to maintain a balance (Fig. 1.4). We will consider the dynamics of the atmosphere and ocean, and the roles played by their circulations in the climate, further below and in Chapters 3 and 5.

## 1.3 Atmosphere and climate

### 1.3.1 Evolution of the atmosphere

The Earth's early atmosphere is believed to have consisted mainly of hydrogen and helium in proportions of about $2 : 1$. This early atmosphere disappeared due to the high surface temperatures that prevailed then, leading to the relatively rapid loss into space of the low molecular weight gases, and to the scavenging action of the vigorous radiation and particle flux from the young Sun in its 'T Tauri' phase, which efficiently removed heavier molecules as well.

Later, a secondary atmosphere was produced as gases were exhaled from the interior of the planet. The composition of this was probably similar to that still being released by modern volcanoes (mostly $H_2O$, with a few percent of $CO_2$ and $SO_2$, plus smaller amounts of CO, S, $Cl_2$, $N_2$, and $H_2$) with unknown, but probably significant, amounts of the cosmologically abundant volatiles $NH_3$ (ammonia) and $CH_4$ (methane) as well. Water vapour condensed as the planet cooled, forming the ocean; oxygen was produced by photosynthesis and ozone by photochemistry, all of these balanced by the continued escape of light gases from the top of the atmosphere, the removal of carbon dioxide by fossil formation, and the loss of oxygen in oxidation processes. The balance is what we see today.

### 1.3.2 Temperature structure

The mean vertical temperature structure of the atmosphere that results from the balance between heating and cooling at various levels is shown in Fig. 1.5. The separation into broad regions by the general pattern of rising and falling temperatures has led to the assignment of names commonly given to these regions, and these are also shown in the figure.

The lowest 10 km or so, in which the temperature drops off with height at a roughly constant rate, is known as the troposphere. 'Tropos' is from the Greek for 'turning', a reference to the importance of convective overturning for transporting heat vertically. It will be shown in Chapter 3 that the so-called *adiabatic* gradient of temperature versus height that is just stable against convection can be calculated simply, and is about $8 \text{ K km}^{-1}$, the exact value depending on the composition, especially the humidity.

At the top of the troposphere (the *tropopause*) the temperature tends to become invariant with height and the atmosphere is stably stratified (i.e. parcels of air tend to move neither up nor down), hence the name *stratosphere* for the overlying region. The key difference between the stratosphere and the troposphere is that, above the tropopause, the lower atmospheric density

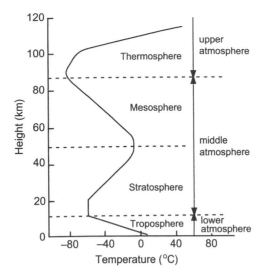

**Fig. 1.5**
The mean vertical temperature profile of the atmosphere, simplified to show the classical naming of the regions or 'spheres'. The interfaces between these, shown as dashed lines, are called, from bottom to top, the tropopause, the stratopause, and the mesopause. These names are still in use, although in modern work it is just as common to refer to the lower, middle, and upper atmosphere, as shown on the right.

means that infrared radiation is more efficient than dynamics at transferring energy vertically, so convection is suppressed. Vertical movement of air does occur in the stratosphere, of course, generally as a result of waves and turbulence, but not as vigorously as it does below the tropopause, and in more limited regions.

Higher in the stratosphere, the temperature starts to rise again, and by the *stratopause* it is almost back to the surface value. The heating that causes this behaviour is largely that due to the absorption of the shorter-wavelength radiation from the Sun, mostly in the UV part of the spectrum, and particularly by ozone. As the density falls further, the absorption is less and the temperature also falls, reaching its lowest value (less than 200K) near an altitude of about 80 km. This region of declining temperature is called the *mesosphere*. Somewhat confusingly, since meso means middle, the stratosphere and mesosphere are now collectively known as the *middle atmosphere*. Along with this, the term *lower atmosphere* is used for the troposphere, and *upper atmosphere* for everything above the mesopause.

The upper atmosphere exhibits very high temperatures because of the energy released during dissociation of atmospheric molecules and atoms by energetic solar photons and solar and cosmic ray particles. Charged particle layers (the *ionosphere*) are also present. Although some interesting physics occurs there, this region is too low in density to be of much significance to the energy budget of the Earth or to the climate at the surface.

### 1.3.3 Pressure, composition, and temperature variations and dynamics

Throughout the lower and middle atmosphere, pressure varies with height according to the familiar laws of hydrostatics.[10] We shall see in Chapter 3 that the hydrostatic equation predicts that pressure falls off exponentially with height in an isothermal atmosphere, and even when the real

[10] To a good approximation, that is; much of the atmosphere is in a state of constant motion and so the laws of hydrostatics do not apply rigorously.

temperature variations described above are taken into account this remains approximately true.

Except for water vapour, the bulk composition of the lower atmosphere is approximately constant everywhere, since the other main constituents (nitrogen, oxygen, carbon dioxide, and argon) are largely chemically inactive and do not condense. Although oxygen is important for the formation of ozone (see Chapter 8), and certain nitrogen compounds are active in the chemical cycle that destroys ozone, the total fraction of N and O atoms involved in these reactions is very small compared to those that make up the bulk of the atmosphere as $N_2$ and $O_2$, so the abundances of the latter remain effectively constant.

The minor constituents of the atmosphere, on the other hand, can be extremely variable and have a very important role in climate physics. Water vapour is a strong absorber of infrared radiation (i.e. a 'greenhouse' gas, see Chapter 7) and of course the level of humidity is a key factor in cloud production and latent heat transfer (Chapter 3). The other key greenhouse gases like carbon dioxide and methane, are also minor constituents of the atmosphere, making up much less than one percent of the total. Some important atmospheric chemistry cycles, like those that form and deplete the stratospheric ozone layer, involve gases that are present only in parts per million or per billion, sometimes even per trillion (Chapter 8).

The atmospheric temperature structure shows variation, not only with height, but also with latitude and season. Figure 1.6 shows the latitude–altitude cross-section of mean temperature at the solstice. Note how the height of the tropopause varies with latitude, being higher in the equatorial regions. It can be found at its lowest, often less than 8 km above the surface, at high latitudes in the winter, and at its highest over Southeast Asia, during the summer monsoon, where it occasionally peaks above 18 km. Since the rate at which temperature declines with height is roughly constant everywhere in the troposphere, this means that the tropopause is coldest in the tropics (Fig. 1.6).

The tropopause is a dynamic surface that can distort and 'fold', bringing ozone-rich stratospheric air into the troposphere, and sometimes all the way to the surface, especially in the northern polar winter. Transfer in the

**Fig. 1.6**

The zonal mean atmospheric temperature (in °C) versus height and latitude, at the time of the solstice (midsummer in one hemisphere, midwinter in the other). The approximate heights of the tropopause, stratopause, and mesopause are indicated by the dashed lines, showing their variation with latitude.

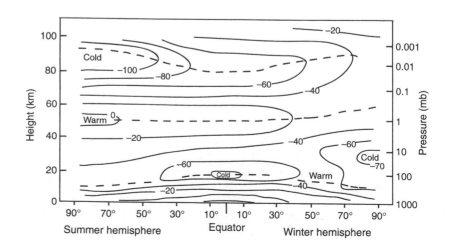

opposite direction occurs mainly in the tropics, which have the highest surface temperatures and the most rapid heating rates, so the transfer of heat and moisture upwards by convection (also discussed in Section 3.5.1) is particularly vigorous there. The resulting thunderstorm activity can break through the tropopause, resulting in the transfer of large parcels of air into the stratosphere. This is probably the principal mechanism for transferring polluted tropospheric air, including ozone-damaging chlorine compounds (see Section 8.5), into the stratosphere.

## 1.4   Ocean and climate

The climate on the Earth would, of course, be very different without the ocean.[11] It stores large quantities of heat and exchanges significant amounts with the air and the land. A vast 'conveyor belt' runs through the global ocean, carrying heat between the warmer and cooler parts of the world. The amount of water vapour in the atmosphere at any time depends primarily on the global balance between evaporation and precipitation over the ocean. The ocean is also a source, sink or reservoir not just for water vapour but also for several other important atmospheric species, carbon dioxide in particular.

[11] As we shall seldom need to refer to individual named oceans when considering the basic physics, we use the term 'ocean' to mean the *global* ocean unless the context makes it clear we mean otherwise.

### 1.4.1   Heat storage and transport

While we experience the effect of the climate mainly through the atmosphere, 71% of the surface of the Earth is covered by water, including 6% of ice cover, and this represents a huge reservoir of heat, as well as moisture, which is available for exchange with the atmosphere. The large heat capacity of the ocean delays the effect of any warming trend in the lower atmosphere: only about 60% of the effect of industrial activity since 1700 is manifest in the surface temperature increases currently being recorded.

The atmosphere and the global ocean are coupled together dynamically, and both have global circulation patterns that have a key role in redistributing heat from the tropics to higher latitudes. Although the ocean moves less rapidly than the atmosphere, its energy storage is much larger. In fact, the top 3.2 m of the ocean has the same heat capacity as the entire atmosphere, and the total ocean heat content is about 1,000 times that of the atmosphere. The net contribution of the ocean to the equator-to-pole-heat flux is about the same as that of the atmosphere, with the ocean dominating in the tropics while the atmosphere takes over at higher latitudes (Fig. 1.7). The combined power involved is about 5 PW ($5 \times 10^{15}$ W), which, as a guide, is equivalent to the output of several million nuclear power stations.

Through its role in reducing the contrast between equatorial and polar temperatures, the ocean provides stability to the climate, but at the same time threatens major changes if its circulation is disturbed. The high salt content of the ocean, and the fact that this varies due to the runoff of fresh water from the land surface, means that the deep ocean circulation is driven by density gradients due to salinity, as well as temperature. The past advance of polar icecaps during the ice ages may have been associated primarily with changes in the ocean circulation and reduced transport of heat to higher latitudes (Chapter 10).

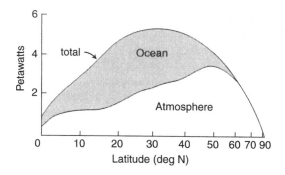

**Fig. 1.7**

The annually averaged poleward transport of heat in the atmosphere and ocean required to balance the solar input in the northern hemisphere. The atmospheric contribution, which is about the same as that of the ocean, is itself in two roughly equal parts: that due to steady flow (winds) and that due to turbulent transport (eddies). Note the non-linear scale of latitude that makes the area between increments of latitude constant. Data from: Estimating the meridional transports in the atmosphere and ocean, B.C. Carissimo, A.H. Oort, and T.H. Vonder Haar, *J. Phys. Oceanog.*, **15**, 82–91, 1985, after Peixoto and Oort, 1992.

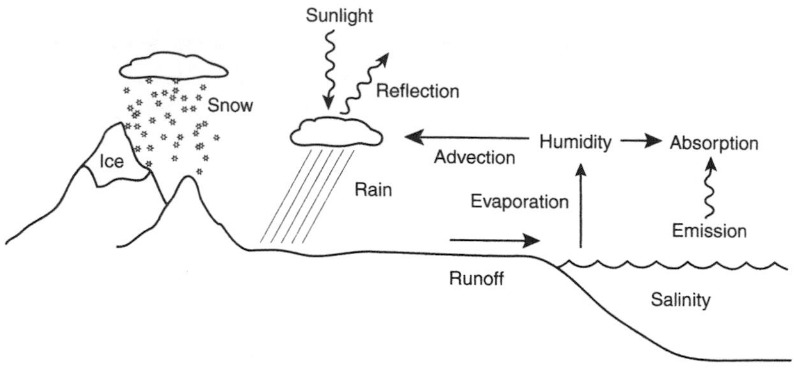

**Fig. 1.8**

A sketch of the hydrological cycle, showing in particular the relationship between the principal reservoir, the ocean, and the formation of clouds, and sources of runoff, which affect the salinity of the ocean and hence its density-driven circulation.

### 1.4.2 Hydrological cycle

The ocean is at the heart of the *hydrological cycle,* holding about 97% of the Earth's water at any given time, some 10 billion cubic kilometres. The transfer of some of this between the ocean, the atmosphere, and the land surface involves the largest movement of any substance on Earth.[12] In addition to precipitation and other aspects of the water budget, the water vapour distribution in the atmosphere affects the Earth's *radiation budget* through the formation of various types of clouds and surface ice cover (Fig. 1.8). Humidity variations in the atmosphere and subsequent cloud formation have their origins principally in evaporation from the oceans, and, conversely, surface freshwater fluxes affect salinity and hence the ocean circulation. The latter in turn affects the storage and transport of heat by the ocean, a major factor in the energetics of the climate system (Section 5.6).

[12] For instance, about 500 trillion tonnes of rain falls on the Earth each year, most of it (about 80%) over the oceans.

## 1.4.3 Carbon dioxide exchange with the oceans

Carbon dioxide is slightly soluble in water, so it tends to get washed out of the atmosphere and into the ocean by the action of rainfall and wave breaking. About one-third of the carbon dioxide released annually from fossil fuels, amounting to about 2 gigatonnes of carbon per year ($2\,\text{GtC yr}^{-1}$ or $2\,\text{PgC yr}^{-1}$),[13] is taken up by the ocean, providing an important reduction in the greenhouse warming caused by the gas.

It is not clear how long the ocean can go on providing this service. It is estimated that the ocean currently contains around 50 times as much carbon dioxide as the atmosphere does; eventually it could become saturated, and the rate of uptake would then slow down. In fact, bearing in mind that the ocean may be warming up, and that $CO_2$ solubility in water decreases with increasing temperature, it could become a net source of $CO_2$ and provide serious levels of positive feedback to global warming trends.

Some of the $CO_2$ in solution in the ocean is retained there permanently, and eventually undergoes sedimentation as a result of chemical and biological processes, forming coral reefs and suchlike. For instance, minute plant life (called *phytoplankton*)[14] floating in the top 200 m of water consumes $CO_2$ during photosynthesis, and converts it to more complex molecules that either end up as food for larger species or sink to the ocean bed at the end of the plankton's short life span (a few days). The rate at which the net removal of $CO_2$ by this 'biological pump' can proceed depends on the reproduction rate of the phytoplankton, which can be very rapid or almost zero, limited by the availability of the other nutrients (since there is plenty of $CO_2$) upon which the organisms depend. These nutrients tend to get depleted in the surface layer, where solar energy is available and where most of the phytoplankton live, but remain abundant at greater depths. Thus, the phytoplankton (and large marine animals like fish that feed on them) tends to thrive at places and times where there is relatively rapid upwelling in the ocean.[15]

Many more measurements are needed before we will know accurately how much $CO_2$ is absorbed annually by the oceans. Equally important is a campaign of measurements to establish how $CO_2$ is subsequently re-released, and in what quantity, where, and when. Ship borne soundings, either by dangling sensors from surface vessels or buoys, or by using submersible manned or robotic craft, remain the only way to obtain data below the surface layer of the oceans, and global and seasonal coverage is obviously a problem. International programmes have been mounted to coordinate the collection of vast amounts of data and organize its analysis. Samples are being collected from many depths using towed, floating and moored sediment traps, pumps, and coring drills; water samples are analysed for their dissolved gas and material content. One of the outcomes is the finding that the rate of exchange of $CO_2$ between ocean and atmosphere in the midlatitude Pacific varies from year to year by as much as a factor of 4. The cause is apparently differences in the temperature structure and the degree of upwelling in the ocean, and the weakening of the winds in the eastern half of the Pacific during 'El Niño' events.

Fluctuations like this obviously add to the difficulty of estimating the overall budget of $CO_2$ accurately. The current best estimates of the global

[13] Climate physics often involves dealing with large numbers like this, and the units adopted are generally those that give researchers conveniently sized digits to deal with. Hence, fluxes in the carbon budget are usually expressed in gigatonnes of carbon per year ($\text{gtC yr}^{-1}$) where giga = billion = $10^9$ and 1 tonne = $10^6$ g. The petagramme or Pg (peta = quadrillion = $10^{15}$) is also sometimes used; 1 Gt = 1 Pg.

[14] Plankton is the collective term for all micro-organisms, plants and animals, in the ocean; phytoplankton refers to the plant component only. These are mostly single-cell algae typically $1-100\,\mu\text{m}$ in size that fix solar energy by photosynthesis, consuming carbon dioxide in the process but requiring dissolved nutrients like nitrates and phosphates to do so.

[15] This behaviour is the basis for the historic 'El Niño' phenomenon, in which Peruvian fishermen noted that their haul was much poorer in some years. We now know that this is the result of a large-scale oscillation in the climate system, which involves major changes in the ocean circulation, and affects the weather world wide (see below, and Section 10.5.3).

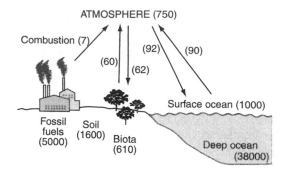

**Fig. 1.9**

The main elements of the carbon cycle, showing the estimated content of the main reservoirs in units of gigatonnes of carbon (GtC), and fluxes, represented by arrows, in GtC yr$^{-1}$. Ocean and living things (biota) remove carbon from the atmosphere, but not fast enough to overcome production by natural and anthropogenic combustion processes. (UK Met Office.)

carbon budget, which includes the carbon dioxide amounts in the atmosphere available to drive greenhouse warming, do not add up unless a large unidentified component, equal to about a third of the total uptake by the ocean each year, is artificially included.

It is crucial to understand this discrepancy, as the balance—presently slightly on the side of removing $CO_2$ from the atmosphere, but likely to go the other way at some point in the future, as noted above—critically depends on it. The most likely place to look for the missing factor is not in a new process altogether, but in the uncertainty in the rate at which known processes, like those involving phytoplankton, absorb atmospheric $CO_2$ into the oceans, and what happens to it after that.

### 1.4.4   Dynamical coupling between the atmosphere and the ocean

The 'El Niño' phenomenon mentioned above is now known to have an importance in climate studies that extends far beyond its long-recognized role as a moderator of the water temperature, and hence the phytoplankton abundance, in the eastern Pacific. More often called the El Niño–Southern Oscillation[16] (ENSO) nowadays, it is the best-studied example of the way the atmosphere and the ocean interact dynamically.

The basis for this interaction involves the way the temperature of the ocean surface affects the rate of heat and moisture transfer to the atmosphere, producing variations in the pressure fields that drive the winds. At the same time, the near-surface wind has a large effect on the motions of the upper layers of the ocean, and on their temperature, creating a feedback loop. This kind of coupling leads to oscillations, with long time scales for those on the largest spatial scales (a few years for ENSO). It is now recognized that resolving this behaviour is crucial for the production of meaningful climate forecasting models on time scales of years to decades (Chapter 11).

### 1.5   Radiative transfer in the atmosphere

Energy is transferred inside the climate system in two principal ways: by motions in the atmosphere and ocean, which transport warm air and water

[16] Both El Niño (in the ocean) and the Southern Oscillation (a regular pressure swing in the atmosphere from Australia to S. America) were phenomena that were noted and described more than a century ago, but it was not until the 1980s that it was realized that they are both aspects of the same resonance in the coupled ocean–atmosphere dynamical system.

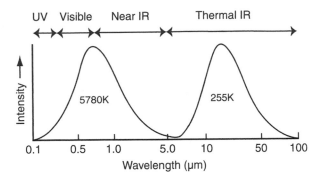

**Fig. 1.10**
The approximate distribution with wavelength of the intensity (energy per unit area per unit wavelength interval) emitted from the Sun and from the Earth, at their mean radiating temperatures of 5780 and 255K, respectively. Here, the plots have been normalized to the same maximum value of intensity[17] (cf. Fig. 2.8).

[17] Radiation is described by flux (the radiant energy crossing unit area), intensity (the flux in unit wavelength or wavenumber interval), and radiance (the intensity in unit solid angle).

to cooler regions and vice versa, and by infrared radiative transfer, this time in the atmosphere only, since the ocean is too opaque to transfer significant amounts of energy as photons. Radiative transfer theory, which is the subject of Chapter 6, can be thought of in two parts, concerned with *solar* (short-wave) and *planetary* (long-wave) radiation respectively.

The main difference is that the former, although it is absorbed and scattered in the atmosphere, originates only in the Sun. Planetary radiation, on the other hand, is emitted from the land, the ocean, and all parts of the atmosphere. The two regimes hardly overlap at all in wavelength: virtually all of the solar energy flux is contained within the range from about 0.2 to about 5.0 μm, which of course includes the *visible* range as well as the UV and the *near infrared*, while the planetary or terrestrial emission occurs between 5.0 and 100 μm, often known as the *thermal infrared* region (Fig. 1.10).

Because the layers in the atmosphere are all emitting and absorbing at the same time, the detailed calculation of the resulting energy budgets and equilibrium temperature profile requires some fairly complicated formulae, the derivation of which will be outlined in Chapter 6. The theory of radiation, based on Planck's radiation law and the laws of radiative transfer, further refines our understanding of sources and sinks of energy in the atmosphere itself and allows us to study simplified concepts such as atmospheres in radiative equilibrium, and to understand the energy budget at the Earth's surface.

Atmospheric absorption in both regimes, and emission in the infrared also, are governed by the principles of molecular spectroscopy that describe the interaction between atmospheric molecules and photons of solar and planetary origin. In the discussion in Chapter 6, it is shown that these laws lead to absorption and emission in spectral *lines*, grouped into *bands*, which absorb and emit only at specific wavelengths that are different for each molecular species. They also explain why minor constituents of the atmosphere like carbon dioxide are so important: they have many bands in the infrared, while the major species $N_2$ and $O_2$, because of their simplicity and symmetry, have none.

A detailed understanding of radiative transfer is also essential to calculate the wavelength distribution of the intensity of the radiation leaving the top of the atmosphere. The details of this distribution depend on the profiles of temperature and composition in the atmosphere below; their measurement and analysis form the basis for the remote sensing techniques, which now dominate Earth system measurements, and without which no quantitative grasp of climate physics would be possible. These are discussed in Chapter 9.

## 1.6  The greenhouse effect

The 'greenhouse' effect is a popular term for the role of the atmosphere in altering the energy balance at the surface of the Earth, raising its mean temperature. The name comes from the analogy with the common garden greenhouse, which transmits heat from the Sun to its interior, mostly as visible radiation, but is opaque to the longer wavelengths at which radiative cooling takes place. (Since the domestic greenhouse also inhibits convective cooling, and insulates against conductive cooling, the analogy should not be taken too far, but the term is too well-ingrained in the popular literature to ignore it altogether).

In the greenhouse effect that warms the surface of the Earth, the panes of glass are replaced by certain gases in the atmosphere. Following re-distribution around the planet, the energy absorbed from the Sun is ventually re-emitted to space. As Earth is a much cooler emitter than the Sun, this re-emission takes place at much longer wavelengths, in the infrared part of the spectrum, as we have seen in section 1.5. While the atmosphere is fairly transparent to visible wavelengths, and so to the incoming sunshine, it is opaque across most of the infrared. Once again, this opacity is due to the absorption bands of minor constituents, especially water vapour, carbon dioxide, methane, and ozone, with virtually no contribution from molecular nitrogen and oxygen. These absorbing or 'greenhouse' gases do not represent all, nor even most of the composition of the atmosphere, but only a very small percentage. However, that percentage is easy to measure, and is indisputably on the increase (Fig. 1.11).

The 'natural' or pre-industrial levels of greenhouse gases (principally water vapour, carbon dioxide, and methane) give rise to an enhancement of the surface temperature of around 30–35K. This means that the human race has depended, throughout its history on the Earth, on the greenhouse effect for its survival, since without it the surface would be frozen everywhere.

Current concern about global warming relates to the possibility of the natural effect being enhanced by up to about 5K in the present century by increased greenhouse gas concentrations of human origin.[18] The increases required to produce such a serious increase in mean surface temperature represent only a small change in the atmospheric composition overall— for example, $CO_2$ might double from 0.035% to 0.07%—and would be otherwise indiscernible.

A great deal of the popular resistance in some quarters to the idea of greenhouse-induced warming as a present-day threat seems to be due to the difficulty many non-scientists have in appreciating that extremely small

[18] With modern measurements it is simply a matter of record that the increases in greenhouse gases are taking place (see Fig. 1.11, for example), but it is less obvious that we can attribute them unequivocally to human tivity, although the ever-increasing release of effluent into the atmosphere can hardly be overlooked. Also, the analysis of gas samples from ice deposits of known age, in glaciers and polar deposits show steady increases from about AD 1800 onwards, while concentrations were relatively constant for at least 1,000 years before that.

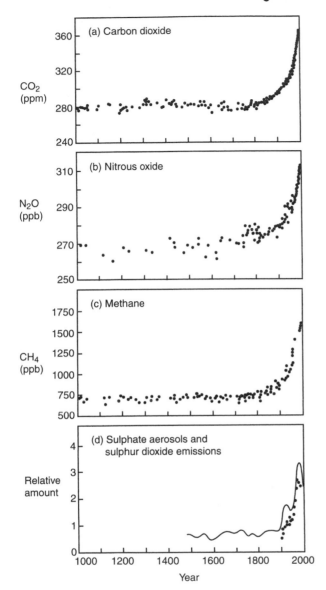

**Fig. 1.11**
The three principal greenhouse gases in the atmosphere, (a) carbon dioxide, (b) nitrous oxide, and (c) methane, are increasing steadily, as a result of processes that occur primarily on the land, including fossil fuel consumption and biomass burning. Although together they are still less than 0.05% of the total atmosphere, each has strong absorption bands in the infrared that alter the energy balance at the surface and tend to induce warming. (d) Sulphate aerosols are also increasing, again mainly due to fossil fuel burning, but in this case they tend to induce global cooling. (Data from the Intergovernmental Panel on Climate Change, IPCC.)

and normally imperceptible changes in atmospheric composition can be responsible. The point that has to be made is that the 'natural' greenhouse effect—which existed before the industrial revolution—is produced by species that are present in parts of less than one in a thousand (some of them much less than that), and that these produce a warming of some 35K without which the Earth would be totally uninhabitable. A change of a few percent in this is enough to have serious consequences.

Of course, the greenhouse effect, like the rest of the climate, does not behave in a linear way, so the fact that the minor constituent inventory of the atmosphere is undoubtedly changing does not mean that the surface temperature will increase in proportion. Some pollutants, in particular sulphur-containing gases like $SO_2$ and $H_2S$ produced from impurities in coal and oil, form aerosols that tend to produce global *cooling* that offsets some of the warming effect of carbon dioxide and methane. It is up to the physicist to determine what will really happen, by acquiring an understanding of processes ranging from the quantum-mechanical behaviour of molecules (which determines the contribution of each species including the relative importance of minor constituents over the abundant ones) to the nature and the effect of the various complex feedback loops, some of which amplify and other of which reduce any tendency for global warming in response to forcing by changes in the 'greenhouse'.

Since it plays no direct role in the global transport of energy, the land surface is in this regard less important than the atmosphere and ocean as a component of the climate system. It is, however, home to all but 0.05% of the Earth's biomass, and is the site of most human activities, and these are important factors determining atmospheric composition and its key role in the atmospheric greenhouse.

As is well known, the oxygen content of the atmosphere is maintained by green plant life, which consumes $CO_2$ and produces oxygen in the process, fuelled by sunlight, known as photosynthesis. Despite conspicuous deforestation and other environmental blight, the proportion of oxygen in the global atmosphere is thought to be stable at the present time, and not therefore a topic that we need to consider at any great length here. The most important thing about plants, for present purposes, is their role in the atmospheric carbon budget; those on land cycle about the same amount of carbon each year as do their marine equivalents, the phytoplankton (Section 1.4.3), in the ocean. The amount of carbon in the atmosphere, primarily as carbon dioxide but also as methane and other gases, is, however, steadily increasing (Fig. 1.11). The cause is believed to be *anthropogenic*, that is, human in origin, with the main culprit the increasing demand for power produced by the combustion of fossil fuels.[19] $CO_2$ and $CH_4$ are two of the most powerful greenhouse gases, because of their strong infrared bands, located in the spectrum at wavelengths that block part of the emission from the surface and lower atmosphere that otherwise would tend to cool the planet. The result is expected to be a general warming trend, and this does seem to be apparent in the global temperature record (Fig. 1.12). We will consider

[19] Of the total carbon dioxide generated by burning fossil fuels, about 20% comes from cars and other forms of transportation, and the rest from static sources, like houses, factories, and power stations. The human population of the world is increasing by about 100 million every year, and the average standard of living—linked to per capita energy consumption—is also increasing.

**Fig. 1.12**
The Earth's mean surface air temperature for the last 120 yrs, corresponding to the period when modern measurements have been made on a more or less global basis. The data are show relative to the average value for the whole period. (Data from IPCC.)

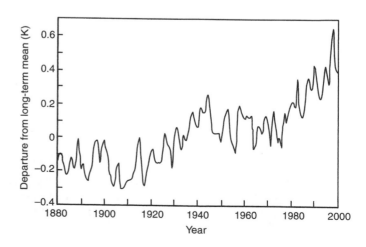

in Chapters 7 and 11 the key question of whether the observed increase is of the right size to match theoretical climate models, and what the latter has to say about future trends.

Between them, the atmospheric minor constituents block direct infrared radiation from the surface to space in all but a few wavelength regions. Thus, only a relatively small part of the cooling of the surface takes place by radiation directly back to space. Instead, the heat is convected upwards, and most of the radiation reaching space leaves the atmosphere at a higher level. The fact that the atmosphere must radiate not only upwards to space, but also downwards, back towards the surface, is the basis for the 'greenhouse' effect that we will discuss in Chapter 7.

## 1.7  The ozone layer and ozone depletion

We have seen that water vapour concentration and the properties of clouds are among the most important variable factors in the climate system, along with atmospheric and oceanic temperatures and the salinity of the oceans. Another key quantity which changes dramatically on a variety of space and time scales is the abundance of photochemically produced species, of which the most important is ozone (triatomic oxygen). This reactive gas is found at all altitudes in the troposphere and stratosphere, but its abundance exhibits a sharp maximum at about 25 km altitude, in the middle of the stratosphere, producing what is commonly known as the ozone 'layer' (Fig. 1.13).

Ozone is a key constituent of the atmosphere because

(1) it is highly chemically reactive and fairly unstable, so its abundance is very variable with height, globally, and seasonally;
(2) it has a large effect on the temperature structure, and the dynamics, of the stratosphere because it absorbs energy from the short-wavelength ultraviolet (UV) end of the solar spectrum;
(3) even the small amount of UV that reaches the surface has a harmful effect on humans (causing sunburn, skin cancer, and eye damage) and plants; without the ozone 'shield' this problem would be much worse;
(4) near-surface ozone is a serious pollutant in cities, where it can reach dangerous levels, but elsewhere tropospheric ozone can be beneficial, by 'scouring' the air of other reactive species.

Figure 1.13 includes a representative vertical profile of atmospheric ozone, showing how the concentration peaks in the stratosphere, with a smaller maximum near the ground. About 90% of the total column abundance of ozone molecules is in the stratospheric ozone layer; this 'good ozone' is the primary shield for the biosphere against solar UV radiation. The 10% in the layer near the surface is an average value, and is much higher in urban areas where we find the 'bad' ozone that has toxic effects on humans and vegetation.

Ozone originates from the dissociation of molecular oxygen into its constituent oxygen atoms in the stratosphere and above by UV solar photons at wavelengths shorter then $0.24\,\mu m$. The liberated O atoms can then attach themselves to remaining $O_2$ molecules to produce $O_3$.

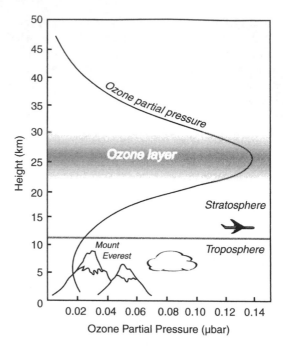

**Fig. 1.13**

A representative vertical profile of atmospheric ozone, with about 90% of the total column abundance of ozone molecules is in the stratospheric ozone layer and 10% in the layer near the surface, mostly concentrated in urban areas.

[20] A catalyst is a substance that enables a chemical reaction without being consumed by it. It can be a solid, a liquid or (as here) an atmospheric gas; only very small quantities are required.

As ozone builds up by this process, it is also being removed. Some wavelengths of UV light dissociate $O_3$ as well as $O_2$, and atomic oxygen O efficiently complicated balance between production and loss in which the ozone abundance ends up being greatest in a relatively narrow vertical range in the stratosphere—the ozone layer.

Using the known rate constants for the reactions and the absorption coefficient in the atmosphere for the dissociating radiation (see Chapter 8) allows the calculation of an expected mean profile for ozone, a task that was first carried out by Chapman in the 1920s. His results were a very poor match to the observed ozone profile, however, being nearly an order of magnitude too high. This remained a major puzzle until the catalytic[20] destruction of ozone was identified in the 1960s, when it was shown that additional processes are involved in the destruction of ozone. The most important of these is a catalytic cycle involving the oxides of nitrogen, NO and $NO_2$, collectively known as $NO_x$, which are present in the stratosphere in tiny amounts, sufficient to have an important effect since the $NO_x$ is continuously recycled.

The $NO_x$ in the stratosphere comes from the breakdown of $N_2O$ (nitrous oxide), which is emitted at the surface as a product of biological activity in the soil and by motor vehicles, or by *in situ* emission of $NO_x$ from aircraft. More recently it has been realized that there are similar cycles involving $HO_x$ and $ClO_x$ (and $BrO_x$), the former being mostly of natural origin, whereas the halogen species result from human activities. An increase in these in recent years, along with the action of stratospheric aerosols of various types, has led to concern about increased ozone destruction, resulting in thinning of the ozone layer and increased penetration of harmful solar UV radiation to the surface.

The most important among the chemically reactive molecules are generally known collectively as 'ozone depleting substances' or ODS. These include chlorofluorocarbons (CFCs), compounds consisting of chlorine, fluorine, and carbon, and hydrochlorofluorocarbons (HCFCs), which also contain hydrogen. Members of both families are commonly used as refrigerants, solvents, and foam blowing agents, while the halons (compounds containing bromine, fluorine, and carbon) are used in fire extinguishers, methyl bromide is used in agriculture, carbon tetrachloride is the original dry cleaning fluid, and methyl chloroform is a widely used organic solvent.

The strong flux of solar UV radiation in the stratosphere, which is responsible for the production of ozone itself, dissociates the ODS molecules, releasing chlorine and bromine which react with ozone in catalytic reactions that restore the halogen atom to repeat the cycle and remove large numbers of ozone molecules from the atmosphere (we discuss the details in Chapter 8). The best-publicized result of ozone depletion is the increased flux of harmful UVB radiation (UVB is the name given to ultraviolet radiation in the 0.28–0.32 μm wavelength range) at the surface, but in climate terms the effect on the temperature structure of the atmosphere is of more concern. Ozone is an important greenhouse gas, as indeed are most of the ODSs as well (see Chapter 8).

The reactions that destroy ozone go faster in the presence of airborne droplets (aerosols), which provide a surface for heterogeneous (multiphase) chemistry, and which also alter the relative abundances of trace species by condensing some of them into the droplets. In Antarctica, the very cold conditions that occur during the winter lead to the formation of hazes of suspended frozen particles known as polar stratospheric clouds or PSCs. The PSCs cause very rapid ozone destruction and produce the notorious Antarctic ozone 'hole' (Fig. 1.14).

## 1.8 Climate observations

As in most branches of Physics, measurements are essential for progress in climate research. In our case, the system under observation is huge,

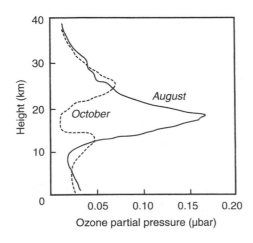

**Fig. 1.14**
Ozone profiles over the Antarctic, before and after the formation of the ozone 'hole'. The hole appears in the Antarctic spring (September) when the polar stratosphere is exposed to sunlight after a long period of continual darkness in the winter, during which polar stratospheric clouds form. (Data from US National Oceanic and Atmospheric Administration, NOAA.)

complex, and time-dependent on a wide range of scales. For many purposes, the measurements must span the globe, as well as the vertical dimension, and must be repeated regularly and sometimes quickly if the shorter timescales are to be captured. For instance, polar ozone destruction is accelerated by the appearance of stratospheric clouds, which can change their size and shape, and even appear and disappear again, on a timescale of hours, associated with transient dynamical phenomena. The prediction of global warming, although less subject to such local, rapid fluctuations, clearly requires global coverage.

To meet these kinds of requirements, the advent of satellite platforms in the 1960s was a godsend, and now much of the most important climate research is conducted using measurements from space. The techniques used mostly involve *remote sensing*—the measurement and analysis of electromagnetic radiation from the region of the atmosphere or ocean under study, as shown symbolically in Fig. 1.15. This radiation may be reflected or scattered sunlight, thermal emission from the atmospheric molecules themselves, or even (in some recent techniques, mostly still in their infancy) backscatter from a laser source on the satellite making the observation.

The instruments used for remote sensing of climate variables like sea and land surface temperature, or the 3-d distributions of greenhouse gases, are themselves the products of sophisticated research projects, and involve the application of the latest in solid-state detector technology, spectral selection methods, and miniaturization. They are usually referred to as either spectrometers (high spectral resolution instruments that can resolve the details in individual spectral bands, as required to study composition, for example, ozone distribution), or radiometers, lower resolution instruments in which the emphasis is on accurate calibration of the intensity of the measured radiation, usually for temperature measurements. Many modern instruments have both attributes simultaneously. Those that operate in the

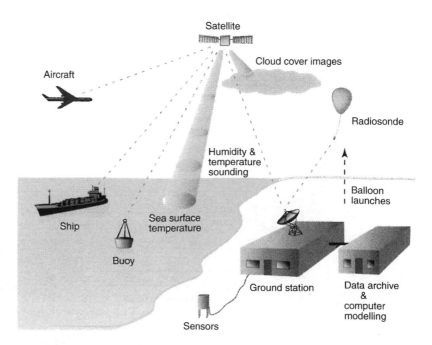

**Fig. 1.15**

An overview of some of the main ways in which climate data are collected. Satellites 'sound' the atmosphere and surface, and also relay data from surface stations, ships, balloons, and aircraft to centres where the data is collected, archived, and made accessible to users.

thermal infrared part of the spectrum are often cooled to low temperatures (typically 80K or less) to increase sensitivity. The science of climate observing instruments will be developed in more detail in Chapter 9.

Radiometric measurements from satellites are not simple to interpret and this has led to another complex sub-field, that of *retrieval theory*. This seeks the best solution for the values of the variables of interest that are consistent with the observed radiation field leaving the top of the atmosphere, and aims to quantify the errors and uncertainties. Essentially, retrieval involves using radiative transfer theory to calculate the spectrum of the outgoing radiation and an inversion technique to find the atmospheric parameters that allow this to match the measurement. Several different approaches are possible, and the fundamentals of these are discussed in Chapter 6.

A key problem in climate modelling is the realistic representation of the time-dependent behaviour of clouds. It is necessary to develop formulae that relate cloud formation processes to their radiative effects in a realistic way. Before this can be attempted on a global scale, models of specific cloudy situations need to be tested with data from experiments so that the way in which the models represent the cloud properties can be optimized. Satellites are excellent for mapping global cloud cover, cloud top height, and optical thickness, but less competent when it comes to determining the crucial physical and microphysical properties of the clouds that have a key role in radiative transfer in cloudy atmospheres. Consequently, much of the current work is being done using instrumented aircraft, which sample the cloud directly and with optical instruments in order to measure its physical thickness, water content, temperature, and pressure characteristics. It is also possible to measure the size (and to some extent the shape, which can be complex if the particles are ice) of cloud droplets by illuminating them with a laser so that their shadows fall onto a detector array. Radiometers on the aircraft can obtain the upward and downward fluxes of radiation above and below the cloud, in carefully selected spectral intervals within the solar and thermal regions, so that these can be compared to fluxes calculated using the measured cloud properties.

Parameterizations that represent large-scale ensembles of cloud, and that therefore are useful in climate models, can then be developed, although it obviously will never be possible to include all of the complexities, particularly since clouds can evolve and change rapidly. Selecting a realistic compromise between accuracy and complexity, especially when it comes to the representation of clouds, is one of the most vexing problems in climate modelling and prediction, and a major source of uncertainty.

## 1.9  The stability of the climate

### 1.9.1  Data on past fluctuations

Measurements from space, and the global coverage that they provide, have only been available to us for a few decades, and scientific climate records of any kind for less than two centuries. Before that, the reconstruction of climatic history relies on so-called paleoclimatic indicators such as tree rings, pollen deposited deep within old bogs, lake and ocean sediments, fossils and other evidence of life forms, and air bubbles trapped in permanent ice fields. Rather remarkably, paleoscientists have been able to construct

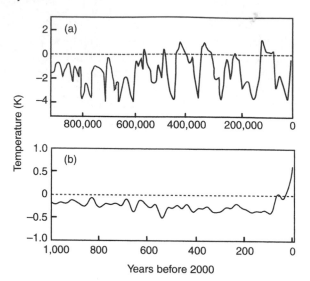

**Fig. 1.16**
Global mean temperatures over (a) the nearly 1,000,000 yrs, and (b) the last 1,000 yrs, relative to the recent (1961–1990) mean global temperature. The older data are estimated from the geological record and are therefore quite uncertain, particularly since global and local effects are hard to separate in sparse data from isolated locations. (Data from IPCC.)

from these a relatively continuous picture of the world's surface temperature over the past one million years (Fig. 1.16).

While treating such data with caution, we can begin to glean some idea of how stable the climate of the Earth is, when subject only to natural perturbations. It is clear that important changes have taken place—the ice ages, for instance, the last of which was around 10,000 yrs ago—that would be disastrous for present-day civilization were they to recur. However, in physical terms, it is remarkable that the climate remains as stable as it does. Models, like those discussed in Chapter 10, suggest that feedback processes involving the atmospheric greenhouse and the surface albedo could, in principle, drive the Earth towards either complete ice coverage or to runaway warming. Yet, over the last million years, the global mean surface temperature has changed by only about 5K, even during the Ice Ages. Dynamical or physical stabilizing mechanisms may be at work within the climate system, including some that have yet to be discovered, or that have been underestimated.[21]

[21] These could possibly include biogeochemical mechanisms, since we have seen that the presence of life on the planet does play a role in stabilizing the climate, for example, by removing $CO_2$ from the atmosphere.

### 1.9.2  Origin of the observed fluctuations

Fluctuations are seen to occur on a wide range of time scales and we can try to attribute them to one or more climate 'drivers'. For instance, the record left in ocean sediments and polar ice fields show variability associated with periods of 19, 23, 41, and 100 thousand years that we now associate with cyclical changes in the Earth's orbit (Section 10.2). As another example, data from the last 20,000 yrs show the rapid onset of the present warm phase following the end of the last ice age, followed by a period when the warming reversed sharply, called the Younger Dryas event (Section 10.5). This was probably associated with changes in the ocean circulation, particularly since the regional data suggests that it was centred on the N. Atlantic. In the seventeenth century, another period of cooling generally known as the 'little ice age' occurred in Europe, but at the same time China was warmer than normal, suggesting a similar origin in ocean circulation changes. The net effect of the 'little ice age' on the best estimate of the global mean temperature was actually quite small, as the lower panel of Fig. 1.16 shows.

Fluctuations in atmospheric composition mean changes in the greenhouse effect and in the ozone layer, upon both of which life on the planet depends. The past record is not very clear, but we do know that both are experiencing measurable, secular changes at the present time due to human activities. We do not really know how the planet is going to respond to the present trends, which provides strong motivation to try to understand climate sensitivity using forecasting models. The largest uncertainty in these is due to the feedback mechanisms that operate to amplify or damp—sometimes it is not clear even after considerable study which will dominate—the effect of an imposed change in the climate system.

Consider, for example, the well-documented increase in atmospheric carbon dioxide that began around the start of the industrial revolution and seems destined to continue for the foreseeable future. On its own, models suggest that this will raise the surface temperature of the Earth by at least 1 K and possibly considerably more in the present century (see Chapter 7). The process has already started, with the following feedbacks occurring as the temperature rises:

1. A general increase in the total water vapour amount in the atmosphere, since the relative humidity (Section 4.4) might be expected to remain approximately constant for moderate temperature changes.
2. Snow cover on land, and ice cover on the sea, melts earlier in the year, reducing the Earth's albedo.
3. The rate at which the ocean absorbs carbon dioxide declines, due to the decrease in the solubility of $CO_2$ in $H_2O$ with increasing temperature. At the same time, the rate of removal of dissolved $CO_2$ may increase, since phytoplankton (and other green plant life, on land as well as in the sea) grow faster at higher temperatures.
4. Changes occur in cloud fractional cover, height, and thickness. Changes in clouds both cool the Earth, by changing its albedo, and warm it, by contributing extra opacity to the greenhouse effect. The net effect depends on the precise optical properties (reflectivity and transmissivity) of the global cloud field, which in turn depend on microphysical properties such as particle size distribution, composition, and shape.
5. Atmospheric and oceanic circulation patterns change, affecting the equator to pole heat flux. The atmosphere and the ocean are coupled, forming a complex dynamical system that has internal fluctuations, both periodic and random, and may have multiple equilibria between which rapid transitions are possible.

## 1.10  Climate modelling

The complex issues raised in the previous section are addressed using mathematical climate models, into which measurements can be introduced or with which they can be compared. For a system as complicated as the climate, models are essential for analysing observations, separating interacting processes so they can be understood, and of course for predicting the future as a function of any change in the boundary conditions. A commonly used yardstick for comparing models is their prediction of how much the mean surface temperature of the Earth will increase as a function of the expected increase in the concentration of greenhouse gases such as $CO_2$. A whole hierarchy of models is in use; we shall deal mainly with simple ones in which the physics is clear and from which the principles underpinning the larger, more complex 'state of the art' climate models may be appreciated.

There are many different types of climate models. Simple models, possibly just a single equation that represents some aspect of the climate system in a very basic way, can be run on a personal computer or even worked out by hand. These usually focus on equilibrium states between one or two specific phenomena while making sweeping assumptions concerning the behaviour of the rest of the system—usually the assumption is that everything not explicitly included as a variable in the model remains unchanged. The relevance of any such model to the real climate is limited, of course, but it does provide useful insights into the role of individual processes.

The model complexity can be gradually increased to include additional processes and the (often subtle) interactions between variables, until the most complete models are arrived at. These can be expected to give the most realistic results, but usually their complexity is such that it is not easy to discern the physics at work, nor to estimate the error bars. Simple models will always have a role to play, not only for education, but to develop the components that are later added to all-inclusive climate models, or to focus on specific features of the climate system (such as cloud formation, or wind-driven coupling between atmosphere and ocean, for example).

The most sophisticated climate models are known as *general circulation models* (GCMs). This name comes from the fact that the earliest GCMs were models that had succeeded in including the fluid dynamical equations that represent the circulation of the atmosphere in a realistic way, a formidable programming challenge for early computers. Nowadays we have ocean GCMs as well, and advanced climate models include atmospheric and oceanic GCMs that are coupled together. The codes they use are often adapted from those used for weather forecasting and consist of a large number of inter-related, time-dependent mathematical equations requiring a massive supercomputer to run. Climate modelling differs from weather forecasting mainly in terms of the time scales involved, and the consequent change in emphasis on different physical processes and parameters. For example, slowly changing factors, such as concentrations of greenhouse gases, remain constant during a weather forecast, but are crucial in a climate model. The deep layers of the oceans, the biosphere, glaciers, and ice-sheets normally change relatively slowly; the solid Earth affects long-term climate through factors like rates of volcanic emission and the location of continents and mountains, which take millions of years to change.

All models, from the simplest to the most complex, make assumptions and simplifications. We are still a long way from being able to make even short-term predictions based on the solution of the laws of physics alone. Climate GCMs, which have to be run for a long time into the future,

generally make compromises in the resolution in space and time represented by the grid on which the equations are solved. This exacts a price
in that sub-grid scale phenomena, such as cloud formation or atmospheric
turbulence, do affect the system as a whole, and have to be parameterized.
Turbulent transfer of heat, a very complex process which is a major factor in
the Earth's energy budget, is sometimes represented by a single parameter
called a diffusion coefficient. In weather forecasts, these parameterization
schemes can be 'tuned' until they give reasonable answers, but in models
designed to mimic future climate change this luxury is not available, and
some uncertainty (we do not even know how much) in the outcome is inevitable. 'Hindcasting'—fitting models to past climate behaviour—can be
helpful in validating models but, as the economists say, past performance
is not necessarily a reliable guide to future behaviour.

Figure 1.17 shows some state-of-the-art climate forecasting, by the
IPCC.[22] It shows model predictions of the sea level rise that is expected to
occur in the present century due to global warming, primarily because of the
melting of polar ice and the thermal expansion of sea water. There are three
main categories of uncertainty in such predictions. The first is the limitation
of the model itself, due to numerical approximations, inexact parameterizations and so forth. The IPCC estimates this uncertainty by repeating
the calculations using six different models, developed quasi-independently
('quasi' because climate scientists talk to each other at national and international meetings, read each others papers, and share ideas) in research
laboratories in several countries. The spread of results between six models
all trying to compute the same thing from the same set of assumptions is
taken to be a measure of the error the mean, and is shown in Fig. 1.17 as
'model predictions'.

[22] IPCC is the Intergovernmental Panel
on Climate Change, a large international
body set up under the auspices of the
World Meteorological Organization to collect together and evaluate information about
climate change and its impact. One of their
most valuable activities has been to compare, combine, and criticize results from
the whole range of climate models from
universities and institutes all over the world.

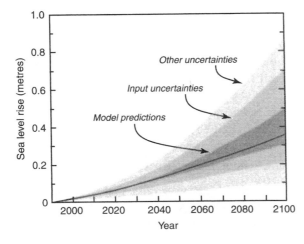

**Fig. 1.17**
Model predictions of the global mean sea level rise expected to occur in the present century due
to global warming. The solid line is the most probable estimate, while the range labelled 'model
predictions' is the envelope of results from six different models. The range labelled 'input
uncertainties' comes from using different emission scenarios for greenhouse gases, while that
labelled 'other uncertainties' further allows for processes that cannot be calculated explicitly at
present, such as land–ice changes, permafrost changes, and sediment deposition.

Next is the uncertainty in the input data itself, in this case the expected rate at which greenhouse gases will build up in the atmosphere during the next 100 yrs. This is dealt with by adopting a large range of possible 'scenarios', ranging from optimistic (strict controls over emissions, new clean technology) through nominal (business as usual) to pessimistic (continued increases as the population of the Earth grows and per capita car ownership and power consumption increases), more than 30 scenarios in all. The range of results corresponding to this is shown in Fig. 1.17 as 'input uncertainties'.

Finally, the outer envelope in Fig. 1.17 reflects the estimated uncertainty due to processes that cannot yet be included in current models, even the latest and largest that represent the state of the art. The largest of these 'other uncertainties' is currently thought to be how fast large ice masses like the Greenland ice sheet (which alone is potentially responsible for around 7 m of global sea level rise if it all melts, which some scenarios say will occur in the next few centuries) will change in response to a general warming. This leaves us with a range of from around 10 to around 90 cm by the year 2100. The former probably can be handled; the latter would be catastrophic (especially as the models go on to predict even more rapid increase in the following century). The case for continuing to improve models and predictions is clear.

For climate forecasts a decade or so ahead, interactions between the atmosphere and the oceans become critically important. This mandates the use of coupled atmosphere–ocean GCMs, and substantially increases the computing time required for a climate modelling experiment, especially since the important ocean currents like the Gulf Stream must be resolved, and long 'spin-up' times are involved before a steady state is reached. The best models also calculate the snow and sea–ice cover, because of their important contribution to the Earth's albedo, and also include chemical and biological processes.

To date, the combined efforts of experimentalists, data analysts, and modellers, has revealed much of the basic physics that produces the present-day climate. We are still a long way from the point where past, let alone future, climatic variations can be reliably explained, nor can we use the models we have developed for the Earth to explain many of the major climate phenomena on the neighbouring Earth-like planets. These failures clearly indicate that the physics in even apparently successful terrestrial climate and weather-forecasting models may be fundamentally deficient.

## 1.11  Climate on other planets

Climate physics is not, of course, a topic that is unique to the Earth. The same physical laws determine the surface environment on the other terrestrial planets with atmospheres, Mars and Venus, and the same gases (principally carbon dioxide and water vapour) are responsible for much of the atmospheric opacity, which gives rise to the greenhouse effect on each planet. Their existence offers an important opportunity to see how the various ideas and models of climate that we shall be exploring work in situations other than that we observe on the Earth.

Venus is a particularly interesting case, where an extreme case of the greenhouse effect raises the surface temperature to 730K (higher than the melting point of lead and tin). This happens in spite of the fact that the net solar input is significantly less than for the Earth, because the very high albedo of Venus (around 0.76, compared to 0.30) more than offsets its greater proximity to the Sun. Venus is, in fact, an example of how clouds increase the greenhouse effect. The sulphuric acid of which the Venusian clouds are composed forms droplets that scatter very conservatively,[23] diffusing a fraction of the incoming sunlight down to the surface. At the same time, they are also very opaque in the infrared. The cloud blanket therefore makes a large contribution to the backwarming effect at the surface, and this, when combined with the contribution of around a million times as much $CO_2$ as the Earth, far outweighs the effect of the planet's high albedo.

The atmosphere of Mars, where the mean surface pressure is only about 7 mb (0.7% of the terrestrial value), is also warmed by airborne particles, in this case windblown dust from the surface. The surface temperature is low by terrestrial standards, about 220K on average, and rises only occasionally above the freezing point of water at equatorial latitudes in the summer. It would, however, be even colder than this were it not for the greenhouse effect produced by the dust. The size of this effect—around 30°C near the surface, and nearly 100°C at the tropopause—emphasizes the potential scale of the aerosol problem on the Earth, and the climatic importance of dust-producing processes like volcanic eruptions and nuclear explosions, particularly if these occur on a global scale.

Mars also possesses the solar system's most dramatic evidence of past climate change. Today's frozen rocks and desert bear unmistakable features of rivers, lakes, and seas. Mars may in the past have had a much thicker atmosphere, much of which now probably lies frozen beneath the surface and in the polar caps. This could have produced a greenhouse effect large enough to raise the temperature and pressure to values that could support liquid water on the surface.

How could Mars have changed so much? The answer may lie in the high eccentricity of its orbit and the large fluctuations in sunfall that result from resonances between this and other orbital parameters. Alternatively, the heating could have been primarily due to volcanic activity that has since subsided. In this case, water vapour, $CO_2$ and other gases, and particles from the interior of Mars could have warmed the planet by greenhouse action and produced lakes and rivers for as long as the volcanoes were sufficiently active.

Recently, images of the polar regions on Mars have provided evidence for changes that are less dramatic but which occur on much more rapid timescales (Fig. 1.18). It remains to be seen whether these are part of ongoing secular changes in the climate of Mars or whether they are part of interannual or other relatively short-term fluctuations. Either way, they represent some of the first direct evidence for climate instabilities on a planet other than the Earth. The intense greenhouse heating on Venus appears to be mainly volcanically driven, and if so is likely to change as the planet's interior cools.

The climates of Mars and Venus are sufficiently like that of the Earth to deserve much closer study. The general message that Earth-like planets clearly can be subject to extremely large and variable greenhouse warmings, for example, is an important and often under-rated one.

[23] 'Conservative' scattering means that an incoming photon that encounters a cloud droplet has its direction changed but is unlikely to be absorbed. Scattering in the Venusian clouds is much more conservative than in terrestrial clouds, because of the difference in composition. This is part of the reason why Venus is such a bright object in the evening sky.

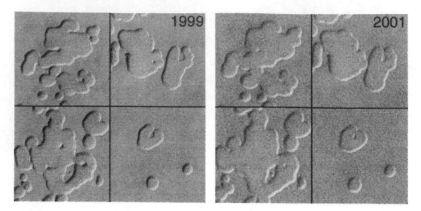

**Fig. 1.18**
Present-day climate variability on Mars as seen from the NASA Mars Global Surveyor spacecraft (Malin Space Science Systems). Close inspection shows that some of these features in the ice cover forming the 'permanent' South polar cap, seen at the same season, have evolved from one Martian year to the next (one Martian year is nearly two Earth years). Such a rapid response, presumable to interannual changes in the seasonal mean temperature, is possible because the cap is made of $CO_2$ ice. Each frame is about 250 m across.

## Further reading

*The Global Climate*, J.T. Houghton (ed.), Cambridge University Press, Cambridge, 1984.
*Principles of Atmospheric Physics and Chemistry*, R.M. Goody, Oxford University Press, New York, 1995.
*Global Physical Climatology*, Dennis L. Hartmann, Academic Press, San Diego, 1994.
*Physics of Climate,* J.P. Peixoto and A. Oort, American Institute of Physics Press, New York, 1992.

## Questions

1. Describe what is meant by 'weather' and 'climate'. What single parameter best summarizes the state of the global climate?
2. What data would you use to define the climate of a specific location, and how might such data be gathered?
3. Summarize the various factors that control the climate of the Earth and comment on how they may have changed over time.
4. Explain the difference between 'climate variability' and 'climate change'.
5. Describe the role of atmospheric gases in energy exchanges in the climate system involving (a) radiation and (b) dynamics.
6. When does the reliable instrument record begin for climate observations? What types of data are used to extend the instrument record further back in time, and what are the limitations of these data for this purpose?
7. Describe the general trend in global temperature from 1880 to the present, and comment on the factors thought to be at work. What do the terms 'little ice age' and 'medieval warm period' refer to? What do they look like in the temperature trend diagram?
8. Why is carbon dioxide the focus of so much attention in climate change studies? Why are much more abundant gases like oxygen not a greater concern?
9. Discuss briefly the likely influence of human activity on the current concentrations of atmospheric trace gases, and on the climate in general.
10. Explain what is meant by the role of feedback in climate change, and give a relevant example.

# Solar radiation and the energy budget of the earth

## 2.1 Sun and climate

Table 2.1 compares the principal sources of energy available to drive the Earth's climate system. It confirms what we would expect intuitively: that virtually all, in fact 99.97%, of the energy arrives from the Sun in the form of electromagnetic radiation. The remainder, from the interior of the Earth, from space in the form of energetic particles (most, but not all, of which also originate in the Sun), cosmic rays, and from human power production using fossil and radioactive fuels, are negligible for most purposes.

This being so, we need to consider how the Sun works as a source of radiation, what is the wavelength distribution of the radiation, and how variable it is with time. Early philosophers thought that the Sun was a solid body, like a huge cannon ball glowing nearly white-hot. In fact, to a useful approximation, the Sun does behave like a hot, solid sphere emitting radiation at a constant temperature of about 6,000K (or, more precisely, 5,780K).

Since we now know that the Sun is a ball of hot, ionized gas or plasma, deviations from this 'rule of thumb' are bound to occur because the radiation it emits into space originates from different depths in the outermost layers, which are at different temperatures. This effect is seen most prominently in the ultraviolet (UV) part of the spectrum, where the Sun emits less strongly than a 5,780K sphere does, which has important consequences for atmospheric chemistry on Earth, for example, the rate of production of atmospheric ozone. At long infrared wavelengths, the emission from the Sun is slightly more than that from a 5,780K sphere, so on average the total energy from a solid body at this temperature is a good match to the total solar output (see Section 2.5 below).

**Table 2.1** The average power at the surface of the Earth for the four main sources of energy entering the climate system

| Source of energy | Average power ($W\,m^{-2}$) |
| --- | --- |
| Solar EM radiation | 240 |
| Energetic particles | 0.001 |
| Geothermal | 0.06 |
| Anthropogenic[24] | 0.02 |

[24] Anthropogenic, from the Greek *anthropos* meaning man, is used in climate studies to refer to anything, for example, air pollution, originating with human beings or in their activities.

## 2.2 Solar physics

The Sun, of course, is a star, and a fairly commonplace one, of the spectral class G2 on the main sequence of the Hertzsprung-Russell diagram. The Sun is larger than the average star, however, in the top 10% by mass of all known stars. Its radius of 696,000 km is a little more than 100 times that of the Earth and about 10 times that of the gas giant planet Jupiter. In mass terms, the Sun weighs in at $2 \times 10^{30}$ kg, which is about a million times more than Jupiter and 300 million Earth masses. According to models of the Sun, the density at the centre is $160,000$ kg m$^{-3}$, or about 15 times that of lead.

Consisting mostly of hydrogen (91.2%) and helium (8.7%), the Sun (like the Earth and the other planets) is believed to be 4.5 billion ($4.5 \times 10^9$) years old. It is likely to last in something like its present form for another 5.5 billion years; in other words our parent star is middle-aged.

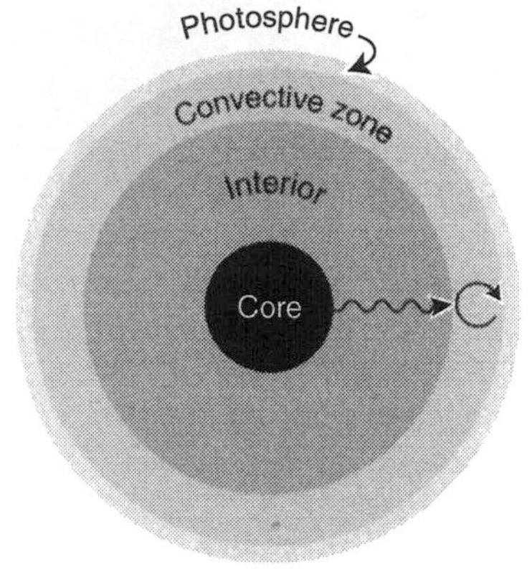

Figure 2.1 shows how the Sun is structured inside. Most of the mass is concentrated in the core, where the central temperature reaches approximately 15,000,000K at a pressure of about 10 Mbar.[25] The Sun rotates with a 25-day period at the solar equator, slowing to around 35 days near the poles.

Of particular interest for climate studies is the temperature of the visible surface or *photosphere*; as already noted this is about 6,000K on average, with fluctuations mostly due to the variable number of sunspots. These are relatively cool regions with temperatures closer to 4,000K. The number of sunspots seen from Earth fluctuates randomly, but also exhibits an internal cycle that yields a global sunspot maximum approximately every 11 yrs.[26] Measurements by satellite instruments (see Fig. 10.5) suggest a variation in the solar 'constant' of 0.08% (or about 1.1 W m$^{-2}$) between the minimum and the maximum of the 11-yr sunspot cycle. What larger fluctuations in solar output occur on timescales of centuries and millennia remains a matter of great debate. Such evidence as exists tentatively suggests a variation between 0.2% and 0.4% over the last 400 yrs.

## 2.3 Source of the Sun's energy

The nature of the Sun's energy source was for centuries the subject of much speculation, and it is still of interest to review some of the more realistic possibilities. Chemical energy is easily shown to be inadequate—if the Sun were made of petrol, and given a supply of oxygen, it would last for around 10,000 yrs at the present rate of heat output.

Gravitational collapse is more plausible: Jupiter emits twice as much energy as it receives from the Sun and the difference is probably due to very slow shrinkage that converts potential to kinetic energy in the interior, the kinetic energy eventually degrading to heat. A simple calculation shows that 20 m yr$^{-1}$ of contraction would produce the required output from the Sun itself, but also shows that could last for only 45 Myrs. This is only 1% of the solar lifetime that has already passed.

Gravitational collapse can, however, raise the interior temperature to the point where thermonuclear fusion takes over. The key reaction is:

$$4(^1\text{H}) \rightarrow 4\text{He} + \text{energy} + 2 \text{ neutrinos}.$$

**Fig. 2.1**
A simple model of the Sun, showing the main radial zones. Most of the energy is produced by nuclear fusion in the core, moves outwards by radiative diffusion in the interior, then is transferred by convection to the photosphere, from where it is emitted to space as radiation.

[25] 1 Mbar = 1 megabar = $10^6$ bars, or 1 million times the mean surface pressure on the Earth. Not to be confused with 1 mbar = 1 millibar = $10^{-3}$ bars.

[26] Sunspots originate in poorly understood instabilities in the Sun's magnetic field, and are linked to a field reversal that coincides with each sunspot maximum, making a 22-year cycle overall.

About 630 million tonnes of hydrogen are converted to 625 tonnes of helium every second. The 5-tonne mass deficit, about 0.7%, when inserted into $E = Mc^2$, gives $4 \times 10^{26}$ W for the power released. The loss to date is then $\sim 5 \times 10^{23}$ tonnes or about 1/4,000 of the solar mass. A somewhat larger amount remains to be converted before the hydrogen in the core is exhausted, corresponding to an estimated lifetime of $10 \times 10^9$ yrs (or 5.5 billion more, as quoted above).

## 2.4  The radiation laws

The radiation laws provide the relationships between the surface temperature of the Sun and the power emitted into space, which must of course match that released by fusion as calculated above. The most important is *Planck's Radiation Law*, which describes the radiant energy (radiance) emitted from a perfectly black object as a function of wavelength for a given temperature.[27] A 'black' object in this sense means one that absorbs all photons that fall on it, that is, a perfect absorber with zero reflectivity. We will discuss the concept and usefulness of black bodies in more detail in later chapters, where we will also make more extensive use of Planck's law, and the other radiation laws, to calculate how the Earth cools to space, and how energy is moved around by radiation within the climate system.

The Planck function is expressed through the following equation for the radiance $R$ as a function of wavelength $\lambda$ and temperature $T$:

$$R(\lambda, T) = \frac{2hc^2}{\lambda^5} \frac{1}{e^{hc/\lambda kT} - 1} \, \mathrm{W\,m^{-3}\,sr^{-1}}$$

where $h$ is Planck's constant ($6.625 \times 10^{-34}$ J s), $k$ is Boltzmann's constant ($1.38 \times 10^{-23}$ J K$^{-1}$), and $c$ is the speed of light, $3 \times 10^8$ m s$^{-1}$. Radiance is the amount of energy per unit time, per unit area, per unit spectral interval, and per unit solid angle, passing through a point. It therefore has units of $\mathrm{J\,s^{-1}\,m^{-2}m^{-1}\,sr^{-1}} = \mathrm{W\,m^{-3}\,sr^{-1}}$ if the spectral interval is measured in wavelength units, that is, metres. However, spectroscopists often like to specify spectral line positions and bandwidths in *wavenumbers*, in units of cm$^{-1}$, instead. The reason is partly historical, but the notation has persisted because the energy of a photon is proportional not to wavelength but to its reciprocal, making wavenumber (the number of wavelengths in one cm) a more intuitive unit for describing photons emitted as a result of transitions between energy levels. (The continued use of cm$^{-1}$ rather than m$^{-1}$ as the unit for wavenumbers is also historical, but does lead to conveniently sized values in the infrared, for instance, $1,000$ cm$^{-1} = 10$ μm). Then radiance has units of $\mathrm{W\,m^{-2}\,(cm^{-1})^{-1}\,sr^{-1}}$, and it is in this form that we often find it in the physics literature.

We will use Planck's expression for radiance in Chapter 6 to calculate radiative transfer within the atmosphere. Two other extremely useful radiation laws, which can be derived from Planck's formula by integration and differentiation respectively, are the Stefan-Boltzmann law for the total power leaving as radiation at all wavelengths from unit area of a surface at temperature $T$, and Wien's law for the wavelength of the maximum rate of

[27] The concept of a blackbody as an object that re-emits all of the radiant energy incident upon it, that is, a perfect emitter and absorber of radiation, was originated by Kirchhoff in 1860. Following his doctoral dissertation at Munich on entropy and the second law of thermodynamics, Planck sought a theoretical explanation of experimental results on how much radiant energy is emitted at different frequencies for a given temperature of the blackbody. The result, which Planck finally reached in 1900, used Boltzmann's insights into the statistical nature of the second law and introduced the concept of the quantum of action, $h$. By comparing the formula to measurements, Planck obtained values for $h$ and for Boltzmann's constant $k$.

emission at a given temperature. The Stefan-Boltzmann law is simply:

$$F = \sigma T^4 \; \mathrm{W\,m^{-2}}$$

where the factor $\sigma = 5.670 \times 10^{-8} \, \mathrm{J\,m^{-2}\,K^{-4}\,s^{-1}}$ is known as Stefan's constant.[28] The quadratic exponent means that doubling the temperature of a body causes the rate at which it emits radiant energy to increase by a factor of 16, more than an order of magnitude. Using the Stefan-Boltzmann formula, we can estimate that the Sun emits about $4 \times 10^{26}$ W.

Wien's 'Displacement' Law states that

$$\lambda_{\max} = \frac{2897}{T} \; \mu\mathrm{m},$$

where $\lambda$ is in $\mu$m and $T$ in K. Thus, a body at a temperature of 6,000K, close to that of the Sun, has its peak emission of energy at $0.5\,\mu$m, in the visible part of the spectrum as we expect since this, of course, is why our eyes have evolved to respond to this wavelength. The temperature of the Earth is nearer to 300K, which would correspond to a peak at $10\,\mu$m, well into the infrared region of the spectrum.

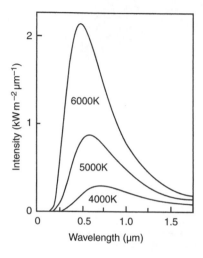

Figure 2.2 shows how emission varies with wavelength, according to Planck's formula, for the approximate temperature of the Sun and for bodies 1,000 and 2,000K cooler. The area under each curve is proportional to $T^4$, as the Stefan-Boltzmann law predicts, while the peak can be seen moving to longer wavelengths as the temperature falls, in accordance with Wien's law. An emitter at the Earth's temperature would, on this plot which has a linear scale, be so close to the zero intensity axis as to be virtually invisible, and its peak would be well off the right-hand side of the diagram (and indeed, right off the page). For comparison, see the log–log plot of these quantities in Fig. 2.8.

**Fig. 2.2**
Plots of the Planck function of intensity (radiance integrated over solid angle in one hemisphere) as a function of wavelength, for the approximate temperature of the surface of the Sun, and two cooler temperatures (sunspot temperatures are typically 4,000–4,500K).

## 2.5  The solar constant

Assuming that the Sun emits uniformly in all directions, we can now obtain a value for the solar constant,[29] defined as the number of watts passing through unit area at the mean distance of the Earth, perpendicular to the Earth–Sun line. Using the more accurate value of $T_S = 5780$K for the effective emitting temperature of the Sun in the Stefan-Boltzmann equation, we multiply $\sigma T_S^4$ by the surface area of the Sun ($\pi R_S^2 = 6.09 \times 10^{12} \, \mathrm{km^2}$) and divide by the surface area of a sphere at the mean distance of Earth's orbit ($\pi R_{SE}^2 = 2.83 \times 10^{17} \, \mathrm{km^2}$). More formally:

$$S = \sigma T_S^4 \left( \frac{R_S}{R_{SE}} \right)^2$$

The calculated value is $S = 1362 \, \mathrm{W\,m^{-2}}$, very close to the mean measured value for the solar constant of $1368 \, \mathrm{W\,m^{-2}}$. Multiplying this by the projected area of the Earth, that is, $\pi R_e^2$ where $R_e$ is the mean radius, we find that about $1.5 \times 10^{17}$ W (150 PW) is intercepted by the planet as a whole, at the top of the atmosphere.

At a point on the surface of the Earth where the Sun is overhead, the radiative energy reaching the surface is equal to the solar constant minus the fraction reflected back to space by clouds etc., and multiplied by the transmission of the atmosphere. If the Sun is at an angle $\theta$ with respect to the zenith, an additional factor of $\cos(\theta)$ is needed to allow for the increase in area illuminated by one solar constant, in addition to factors that allow for any reflection or absorption. Thus we can write

$$F = S(1 - R)T \cos \theta \ \mathrm{W\,m^{-2}}$$

where $R$ and $T$ are effective reflection and transmission terms. These factors are variable with time, of course, and since reflection and absorption of radiation takes place over a range of altitudes, quite complicated to calculate in any given situation (see Chapter 6). Measurements of the solar constant make sense only if obtained from platforms in space, above the atmosphere, facing the Sun. Such observations are summarized in Chapter 10.

## 2.6  The solar spectrum

We have seen that the total energy emitted by the Sun is about the same as that of a blackbody at a temperature of 5,780K; the distribution of the solar intensity with wavelength is also quite close to that of a blackbody at this temperature (Fig. 2.3).[30] The observed deviations from this are most significant in the UV region, where the molecular bands responsible for ozone production are found (Section 8.2). These deviations occur because Planck's formula assumes that the radiation is emitted from a body at a single, uniform temperature, whereas the temperature of the Sun increases with depth, and the emission to space occurs over a range of depths. The depth from which the emission comes varies somewhat with wavelength, due to absorption processes that occur within the Sun, for example, in the Fraunhofer lines of highly ionized metals as Mg and Al.

When measured above the atmosphere, about 9% of the solar energy is in the form of ultraviolet radiation ($\lambda \leq 0.40\,\mu\mathrm{m}$), approximately 38%

[30] The shape of the spectrum, together with the fact that the eye has maximum sensitivity near $0.55\,\mu\mathrm{m}$, accounts for the characteristic yellow–white colour of the Sun.

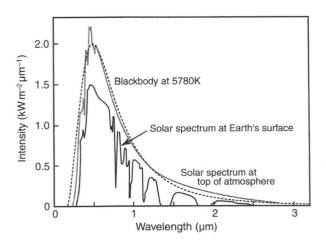

**Fig. 2.3**
The upper and lower solid curves show the observed intensity versus wavelength—the solar spectrum—at the top of the Earth's atmosphere and at the surface. The ragged nature of each is due to the energy removed by absorption bands in the solar photosphere and the Earth's atmosphere respectively. The smooth dashed line is the spectrum the Sun would have if it were a blackbody at the 'best fit' temperature of 5780K.

is visible ($0.40 \leq \lambda \leq 0.70$ μm) and about 53% lies in the near infrared ($0.7 \leq \lambda \leq 4.0$ μm). The sun also emits shorter wavelength X-rays and gamma rays, and longer wavelength thermal infrared ($4.0 \leq \lambda \leq 100$ μm), microwave ($0.1 \leq \lambda \leq 10$ mm) and radio ($1$ cm $\leq \lambda$) photons, but these make negligible contributions to the total solar energy input to the Earth.

The integrated energy from the Sun fluctuates slightly with time, with amplitude of the order of 0.1%, but the intensity at some wavelengths varies much more than this, as a function of conditions on the Sun. All of these variations can be of considerable interest since the response of the Earth is difficult to predict and, if it can be observed, quite informative about the Earth–Sun relationship. For example, it has long been discussed whether weather conditions at the surface are influenced by short-term solar variability or not, and the debate continues.

## 2.7  Solar observations

Some of the best series of measurements of the total solar input are from the active cavity radiometer irradiance monitor (ACRIM) series of instruments, which have flow on a variety of spacecraft since the Solar Maximum Mission in 1980, and are planned to continue. The basic principle is to maintain the temperature of a blackbody, in the form of a blackened cavity, at a constant value by electrical heating while a shutter over the entrance aperture facing the Sun is opened and closed. The thermodynamic state of the cavity remains constant and a measurement of the current required to compensate for the removal of solar heating allows the latter to be deduced with high accuracy. ACRIM measurements appear as part of Fig. 10.5 in the section on solar variability.

Measurements of the solar spectrum from space have been made by instruments known as SUSIM (Solar UV Spectral Irradiance Monitor) and SOLSTICE (Solar Stellar Irradiance Comparison Experiment) flying on NASA's Upper Atmosphere Research Satellite (UARS) launched in 1991. These are basically grating spectrometers, which, between them, cover the range from 119 to 420 nm in the UV (Fig. 2.4). The solar variability is greatest here (of the order of 10% at some wavelengths) and these are the more energetic photons that drive photochemical reactions in the atmosphere. Since the absolute intensity as a function of wavelength is

**Fig. 2.4**

The solar UV spectrum as measured on 3 August 2004 by the SUSIM instrument. The prominent line at 0.122 μm is the Lyman alpha line of hydrogen.

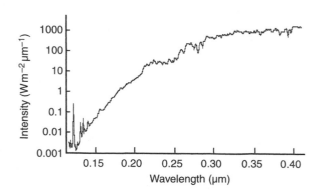

required, calibration is a major issue; SUSIM uses a series of standard lamps, while SOLSTICE has an innovative approach by which the input aperture is variable, allowing the observation of a number of standard UV stars.

The results from instruments of this type show, as might be expected, large variations, especially in the bright Lyman alpha band, with the same frequency as the solar rotation and with sunspot activity, although other, smaller, periodic and episodic variations are also seen. The 27-day cycle produces a corresponding signal in the atmospheric temperature at very high altitudes, and sunspot eruptions on the Sun produce particle fluxes that have been observed to strongly modulate the nitric oxide abundance in the thermosphere, over localized regions and for a short time. This affects the abundance of ozone in the upper stratosphere but it is not yet clear whether there is any measurable influence of solar fluctuations on the lower atmosphere.

## 2.8 Absorption of solar radiation in the atmosphere

Figure 2.5 shows how four common molecules contribute to atmospheric absorption, as a function of wavelength, and the solar spectrum after passage through the Earth's atmosphere to the surface. At some wavelengths, the flux of energy reaches the surface almost unattenuated; at others, it is removed virtually entirely by absorption in the spectral bands of atmospheric gases. Ozone absorbs strongly in the UV, including the so-called Hartley and Huggins bands that are primarily responsible for heating the atmosphere at altitudes around 50 km, the visible (Chappuis) band at 0.6 $\mu$m, and in the infrared at 9.6 $\mu$m (see Section 8.2). Molecular oxygen contributes in the UV (the bands that give rise to ozone production). Water vapour, $H_2O$, absorbs strongly at 6.3 $\mu$m, and in the far infrared beyond 20 $\mu$m. Carbon dioxide, $CO_2$, has many bands in the near infrared, plus two in the mid infrared at 4.3 and 15 $\mu$m that are particularly strong. The 15 $\mu$m band falls near the peak of the Planck function for typical atmospheric temperatures. In addition to these, there are contributions by $CH_4$, $CO$, $N_2O$,

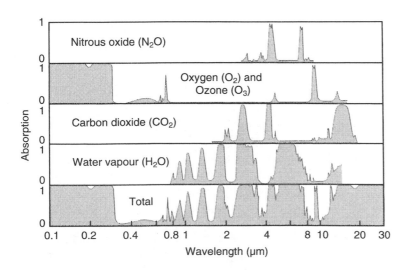

**Fig. 2.5**

Absorption spectra of atmospheric gases, for a vertical column from the surface to space. From the top: nitrous oxide; ozone (with a contribution from ordinary oxygen, $O_2$, in the UV); carbon dioxide; water vapour; and the sum of all species in the bottom panel. Regions where the absorption is low, such as those in the visible part of the spectrum, or near 8 and 12 $\mu$m in the infrared, are called spectral 'windows'.

NO, etc. at various wavelengths and *continuum* absorption (i.e. not restricted to specific bands) by aerosols and cloud particles. To understand the origin and behaviour of these bands requires a quantum-mechanical treatment that will be further discussed in Chapter 6.

## 2.9 The balance between incoming solar and outgoing thermal radiation

About one third (30%) of the total energy from the Sun that reaches the Earth is reflected back into space, mainly by clouds (Fig. 2.6). About half (50%) reaches the surface, where it is absorbed and produces heating. Gases, clouds, and aerosols at various levels in the atmosphere absorb the remaining 20%.

We have seen that 1,368 W m$^{-2}$ falls on a sphere with the radius of the Earth's orbit. Since the actual surface area of the planet is 4 times its cross-sectional area (see Fig. 7.3), the average energy into the climate system is 70% of 1,368/4 or 239 W m$^{-2}$. The value at any given location depends, of course, on latitude, solar zenith angle (time of day), and atmospheric and surface conditions, especially factors like cloud and ice cover that strongly affect the albedo. Figure 2.7 shows how the foreshortening effect of the curvature of the Earth affects the annual mean sunfall (energy per unit area) at the surface, for the hypothetical case where the Earth's axial tilt or obliquity (angle between the equator of the planet and the line from the centre of the Earth to the centre of the Sun) is zero, and also for the real value of obliquity = 23.5°. Note that, if the Earth's spin axis were perpendicular to its orbit, neither pole would receive any direct heating from the Sun. In reality, each is tilted towards the Sun during the relevant summer season and the distribution with latitude of mean sunfall is less extreme, as the figure shows. There is still a difference of more than a factor of two between equator and pole, that would correspond to a much more extreme gradient in climate with latitude, were it not for the effect of the oceans and atmosphere in transferring heat polewards.

Since most of the surface is covered with water (about 70%), it is interesting to calculate what would happen if the entire solar radiation incident on the Earth went into heating the ocean. Making the simplifying assumption that the ocean is well-mixed, and so all at temperature $T$, the rate at which $T$ would change with time $t$ is

$$\frac{\mathrm{d}T}{\mathrm{d}t} mC = 239 \text{ W m}^2,$$

where $m$ is the mass of a column of water (in kg m$^{-2}$) and $C$ is the heat capacity per unit mass ($\sim$4.2 kJ kg$^{-1}$ K$^{-1}$).

The mean depth of the ocean, averaged over the globe, is about 4 km, and the density of water is $\sim$1,000 kg m$^{-3}$, so

$$\frac{\mathrm{d}T}{\mathrm{d}t} \approx 0.5 \text{ K yr}^{-1}.$$

In fact, only about the top 100 m of the ocean is well-mixed (Section 5.4), and since all of the solar energy is absorbed near the surface, a more realistic

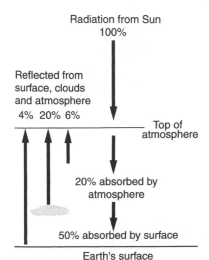

**Fig. 2.6**
This diagram summarizes what happens to solar radiation after it falls on the Earth. Roughly half reaches the surface; nearly a third is reflected back to space, mainly from clouds, and the rest is absorbed by atmospheric gases (see Fig. 2.5).

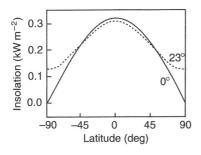

**Fig. 2.7**
The insolation (solar energy per unit area) as a function of latitude, averaged over the year, for the case if the Earth's spin axis was perpendicular to the plane of its orbit around the Sun (obliquity = 0°, solid curve) and for the actual obliquity of 23.5° (dashed curve).

estimate for the heating rate would be

$$\frac{dT}{dt} \approx 20K \ yr^{-1}.$$

In fact, the observed value for $dT/dt$ is less than about $0.01K \ yr^{-1}$. The reason for the difference is, of course, that the ocean, and the Earth as a whole, is not only heated by radiation but also cools radiatively as well, the energy being ultimately radiated back into space. Still focussing on the ocean, the cooling is in part *directly* to space, especially in the spectral 'windows' (regions of low absorption, as shown in the bottom panel of Fig. 2.5), and partly to the atmosphere, to be transported upwards dynamically and then radiated away.

Overall, the Earth must radiate away the same amount of energy as it receives, since its mean temperature is at least approximately constant with time. According to Planck's formula and Wien's law, the terrestrial emission is expected to occur at longer wavelengths than that of the much hotter Sun (Fig. 2.8). According to Stefan's Law, the total amount emitted (the area under the curve) by a 6,000K body (approximately the temperature of the Sun) is greater than that by a 300K body (approximately the surface temperature of the Earth) by a factor of $(6,000/300)^4 = 160,000$. Since the planet is in overall radiative balance, this factor must be approximately equal to the reciprocal of the fraction of solar energy intercepted by the Earth's surface:

$$\frac{(1-A)}{4}\left(\frac{R_S}{R_{ES}}\right)^2 = \frac{0.7}{4}\left(\frac{700,000}{150,000,000}\right)^2 = 3.8 \times 10^{-6},$$

where the factor $1/4$ again allows for the ratio of the projected area to the actual surface area of the Earth, and $A = 0.3$ for the albedo, the fraction of the solar flux reflected back to space.

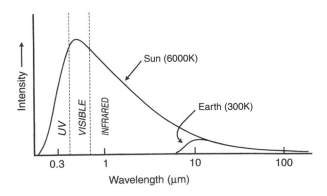

**Fig. 2.8**
A schematic of the emission spectra of Planck's law emitters (black bodies) at uniform temperatures of 6,000 and 300K, roughly those of the Sun and Earth respectively. Both axes are logarithmic; the ratio of the total power emitted by unit area of the Sun, integrated over all wavelengths, to the same quantity for the Earth, is about 160,000.

[31] This kind of *energy balance model* is useful in the study of the greenhouse effect—see Chapter 7.

Since the reciprocal of 160,000 is $6.25 \times 10^{-6}$, these calculations do not in fact seem to indicate overall energy balance for the Earth. The reason, of course, is that the mean surface temperature of around 300K (actually 288K) is not the effective temperature at which the Earth emits radiation to space: we have to allow for the atmosphere that overlies the surface, and which is colder and partially opaque. If we re-do the calculation above assuming the Earth to be in overall radiative energy balance[31] we obtain a value for the actual effective emitting temperature of the Earth and find that it is closer to 255K. The origin of this value can be appreciated by looking at the measured emission spectrum of the Earth as seen from space, shown in Fig. 2.9. The area under the terrestrial emission curve is about the same, when averaged over the globe and over the seasons, as that under the curve calculated from the Planck formula for a temperature of 255K.

Thus we arrive at the concept shown in Fig. 2.10. Here, the solar spectrum is represented by a black body at 5,780K, attenuated by the factor $((1 - A)/4) \times (R_S/R_{ES})^2$ as before to allow for the inverse square law, the spherical nature of the Earth and its albedo, $A$. The Earth is represented by a black body at 255K. We note that the two curves hardly overlap at all; for the Earth, absorption of energy and its subsequent re-emission take place in almost totally separate wavelength regimes, the UV, visible, and near infrared (or 'solar') in the former case, and the middle and far (or 'thermal') infrared in the latter. The boundary between the two regimes falls at about 5 μm. The integrated energy per second absorbed and emitted, represented by the areas under the curves, is equal, representing the overall radiative energy balance of the planet.

Superimposed on the equivalent blackbody curves in Fig. 2.10 is the absorption profile of the atmosphere, from Fig. 2.5, to show approximately where the strong absorption bands of the infrared-active gases lie on the

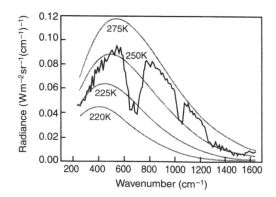

**Fig. 2.9**

The spectrum of the infrared energy emitted by the Earth.[32] The various features are the absorption/emission bands of atmospheric gases, especially water vapour, ozone, and carbon dioxide (Fig. 2.5). The area under the Earth's spectrum, when averaged over latitude, longitude, and time, and integrated over wavelength, is about the same as the area obtained by integrating the Planck function (represented at four different temperatures by the smooth curves) for a temperature of 255K. At this temperature, the thermal infrared emission from the Earth just balances the incoming solar radiative energy at shorter UV, visible, and near-infrared wavelengths.

[32] This spectrum is in wavenumber units: 1 cm$^{-1}$ means one wavelength per cm, so 1,000 cm$^{-1}$ is the same as a wavelength of 10 μm (= $10^3$ cm).

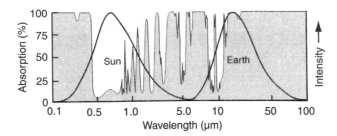

**Fig. 2.10**
To illustrate the radiative energy balance of the Earth, this Figure shows the solar spectrum, represented by a 5,780K black body scaled down by factors that allow for the distance from the surface of the Sun to the Earth, the sphericity of the Earth, and its albedo; and the Earth's emission represented by a 255K blackbody. Now, the area under the curves is equal, since the Earth is in radiative balance with the Sun overall. A representative absorption spectrum of the Earth's atmosphere, for a vertical column from the surface to space, is also shown (left-hand scale).

wavelength axis. A key point to note that there is considerable more opacity overall in the thermal regime than in the solar, the basic factor responsible for the greenhouse effect, a key regulator of the climate that we will consider in detail in Chapter 7.

## Further reading

*Lectures on Solar Physics*, H.M.M. Antia, P. Ulmschneider, A. Bhatnagar, and H.M. Antia, Springer-Verlag, New York, 2003.
*The Sun, An Introduction*, M. Stix, 2nd edition, Springer-Verlag, Berlin, 2002.
*Astrophysics of the Sun*, H. Zirin, Cambridge University Press, Cambridge, 1998.
For discussions and references about solar radiation in the Earth's atmosphere, see Chapter 6.

## Questions

1. What is meant by the *equivalent blackbody* or *effective emitting* temperature of the Sun? Describe the Stefan-Boltzmann radiation law and use it to calculate the power of the total radiant energy emitted by the Sun, assuming an effective emitting temperature of 5,780K.
2. Using the result of the previous question, calculate the amount of mass lost by the Sun each second to generate the energy emitted as radiation. How long would the Sun continue to burn if all of its mass could be converted to energy at the same rate? What is the actual expected lifetime of the Sun, and why is this different from the number you just calculated?
3. Define the term *solar constant*. To what extent is it actually constant? What are the two principal factors that relate the solar flux at a point on the Earth's surface to the solar constant?
4. Assuming the Sun has a radius of 70,000 km and an effective mean temperature of 5,780K, obtain an expression for the effective emitting temperature $T_E$ of the Earth, that is, the temperature at which the Earth would be in equilibrium with the Sun if both behaved as blackbodies. Assuming that the mean Earth–Sun distance is 150 million km, and that the Earth has an albedo of 0.3, show that $T_E = 255\,K$. Why does this differ from the surface temperature of the Earth?
5. Estimate the rate at which the Earth's surface would cool if the Sun stopped shining, stating any assumptions made.

# 3 Atmosphere and climate

## 3.1 Introduction

The atmosphere is a complicated thermodynamic 'heat engine', consisting of a layer of gas, vertically and to some extent horizontally inhomogeneous, gravitationally bound to the surface of a rapidly rotating[33] sphere with complex surface topography, and fed with radiant energy originating effectively at infinity. Its response is extremely variable on time-scales of hours or days, as our weather frequently testifies. In order to deal quantitatively with the role of the atmosphere in the climate, we want to investigate the *mean* state of this complex and dynamic system, typically averaged over a year or longer so that fluctuations due to weather (and even those due to the seasons in many cases) are suppressed in favour of *secular* or long-term changes, particularly those which may occur in response to external forcing.

Fortunately, this long-term outlook allows us to make major simplifications, at least when trying to understand the key processes important for climate at a basic level. It is often useful to consider a one-dimensional representation of the atmosphere, where the variable is height or pressure. This means we represent the whole atmosphere by a single vertical profile that is typical in some way, either a global mean, or a representative mid-latitude profile, possibly averaged around a circle of longitude at that latitude.

One-dimentional models are only useful up to a point, however. Some problems at the heart of current concern over climate change, for example, increased ice cover at the poles due to reduced ocean circulation from the tropics, affect the mean temperature of the whole planet, but the physics are lost in a 1-d model. The same is true of the stratospheric ozone layer, where the main source is at the equator, the biggest losses are at the south pole in winter, and transport has a crucial role. In the latter case, a 2-d model, where the two dimensions are height and latitude, can be useful. Because of the shortness of the diurnal cycle relative to the seasons, the atmospheric climate variables such as temperature and composition vary much less with longitude (in the *zonal* direction) than with latitude (in the equator-to-pole or *meridional* direction).

Not all processes in the climate system are zonally symmetric, and this is particularly true of the oceans where there are physical barriers in the form of the continents to take into account. We need sometimes, therefore, to think in all three spatial dimensions. Fully resolved three dimensional models, often called general circulation models or GCMs (see Chapter 11)

[33] One rotation in 24 h may not seem very rapid, but it corresponds to a velocity of about 500 ms$^{-1}$ at the equator of the planet. This is large enough for the global-scale atmospheric motions to be dominated by the *Coriolis force* (see Section 5.6.1), which is the criterion for deciding whether rotation is 'rapid'.

are indispensable once we move beyond understanding specific processes to trying to simulate and forecast the behaviour of the complete climate system. We will look at the predictions of these complex models later; in the present chapter we focus on the basic physics and processes that are common to all models, from the simplest to the most sophisticated.

## 3.2 Atmospheric composition

The present day gross composition of the Earth's atmosphere is shown in Table 3.1. Nitrogen and oxygen are commonly referred to as *major* constituents; water vapour, argon, and carbon dioxide as *minor* constituents, and the others as *trace* constituents. The major constituents $N_2$ and $O_2$, and the noble gases Ar, Ne, He, Kr, and Xe, are generally chemically stable and transparent to both solar and terrestrial radiation, so it is often the minor gases $H_2O$ and $CO_2$, and the trace gases like $CH_4$, $N_2O$, and $O_3$, that we find playing a key role in many climate processes, despite their small relative abundances.

Water, as vapour and as clouds, and ozone are both highly variable, and have particularly important interactions with atmospheric radiation, making them 'wild cards' in the climate system that require particularly detailed study.

## 3.3 Units of pressure

The pressure at any point on the surface (or above it) is simply the mass of the overlying air multiplied by $g$, the acceleration due to gravity at that point. 'Standard' pressure at the surface was originally defined as 760 mm Hg, which is known as 1 atmosphere (atm). Unfortunately, a slightly different definition is also current, in which the standard is the *bar*, equal to $10^5$ N m$^{-2}$ or pascals (Pa). Thus, 1 atm $=$ 1.013 bar. A convenient and frequently used unit is the millibar (mbar) or hectopascal (hPa, $=$ 100 Pa),

**Table 3.1** Mean composition of dry air (% by volume) in the troposphere. The water vapour ($H_2O$) content varies from almost zero to several 4%, with an average value of about 0.8%. Tropospheric ozone is also highly variable. The '$>$' next to the values for carbon dioxide $CO_2$, methane $CH_4$, and nitrous oxide $N_2O$, indicates that they are all steadily increasing.

| | |
|---|---|
| $N_2$ | 78.08 |
| $O_2$ | 20.95 |
| $CO_2$ | $>0.033$ |
| Ar | 0.934 |
| Ne | $1.82\times10^{-3}$ |
| He | $5.24\times10^{-4}$ |
| Kr | $1.14\times10^{-4}$ |
| Xe | $8.7\times10^{-6}$ |
| $H_2$ | $5.0\times10^{-5}$ |
| $CH_4$ | $>2.0\times10^{-4}$ |
| $N_2O$ | $>5.0\times10^{-5}$ |
| $O_3$ | $\sim4.0\times10^{-6}$ |

[34] Because of the interdisciplinary background of climate physics, we will
encounter several situations where different systems of units are in common use
to describe the same thing. Until complete standardisation is achieved, the student needs to become familiar with all of
them, and to take care when using mixed
units in calculations.

both equal to 1/1,000 bar or 1/1,013 atm. To complete this feast of units[34]
we note that mean surface pressure is also approximately equivalent to
$1 \, \text{kg cm}^{-2}$, which in old units is $\sim 15 \, \text{lb in.}^{-2}$ or about 1 ton per square foot.

## 3.4　The variation of pressure with height

Pressure is observed to fall monotonically with height, with a nearly linear
relationship between log pressure and altitude (Fig. 3.1). If motions are
neglected, this simply reflects the balance between the pressure gradient
and gravity, and the relationship between pressure and height is governed
by the *hydrostatic equation*:

$$\mathrm{d}p = -\rho g \, \mathrm{d}z,$$

where $p$ is pressure, $\rho$ is density, and $z$ is height. The expression on the right
is just the mass of a column of air of unit cross-sectional area and height
$\mathrm{d}z$, multiplied by the acceleration due to gravity, $g$. The difference in force
per unit area between the top and bottom of the column of height $\mathrm{d}z$ is the
pressure exerted by this mass of air, $\mathrm{d}p$, with the minus sign arising because
of the convention that $p$ decreases in the direction in which $z$ increases, that
is, upwards (Fig. 3.2).

The ideal gas law for 1 mole is

$$pV = RT$$

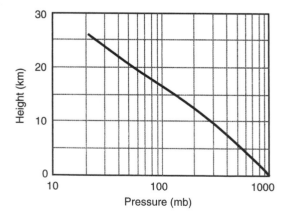

**Fig. 3.1**
Measurements from a balloon of the
atmospheric pressure as a function of
height at one particular place (2°S, 169°W)
and time (1200 on 15 March 1993). With
the use of a linear height scale and a log
pressure scale, the plot is nearly a straight
line (and would be perfectly straight if the
atmosphere were static and isothermal).

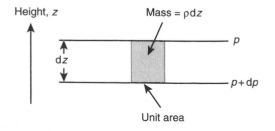

**Fig. 3.2**
The mass of a column of air between
surfaces at pressure $p$ and $p + \mathrm{d}p$,
separated by height $\mathrm{d}z$, is just the density
$\rho$ times $\mathrm{d}z$, so its contribution to the
pressure (force per unit area) is
$\mathrm{d}p = -\rho g \, \mathrm{d}z$.

and density $\rho$ is equal to $M/V$, so

$$\rho = \frac{Mp}{RT},$$

where $M$ is the mass of 1 mole and $R$ is the gas constant.

Substituting from the hydrostatic equation:

$$\frac{\mathrm{d}p}{p} = -\frac{\mathrm{d}z}{H},$$

where $H = \frac{RT}{mg}$ and $H$ is called the *scale height*. The physical meaning of $H$ is that it is the height over which the pressure drops by $1/e$,[35] as we can see by integrating from the surface ($z = 0$) to height $z$:

$$p = p_0 \exp\left(-\int_0^z \frac{\mathrm{d}z}{H}\right).$$

[35] The pressure at the top of a column of height $H$ (e.g. at the summit of Mt Everest, which, at 8.85 km, is approximately one scale height above sea level) is about one-third of that at the bottom.

$H$ varies throughout the atmosphere because of the dependence on temperature. However, $H$ is often taken to be constant in approximate work, especially over a limited height range where $T$ does not vary much. Even over the whole atmosphere, $T$ and therefore $H$ varies by only about 30%, since 200K $< T <$ 300K at most levels. Given that $R = 8.314\,\mathrm{J\,K^{-1}\,mole^{-1}}$ and $M = 28.96 \times 10^{-3}\,\mathrm{kg\,mole^{-1}}$ for dry air, we find $H = 8.45$ km at 290K (typical of ground level) while $H$ drops to 5.84 km in the upper mesosphere ($T = 200$K).

The *thickness* of a layer between pressures $p_1$ and $p_2$ is given by

$$\Delta_z = -\frac{RT}{Mg} \log\left(\frac{p_1}{p_2}\right),$$

where $T$ is now the mean temperature of the layer. This equation is the basis of the pressure altimeters used in aircraft. Note however that the determination of absolute altitude requires knowledge (or an estimate) of the temperature profile and the surface pressure, as well as the humidity, which is required in order to obtain $M$ (see Section 4.4).

Because the hydrostatic equation describes a one-to-one relationship between pressure and height, the two are often used interchangeably. Height is often used when discussing measurements and observations while pressure (or log pressure) is found to be the more convenient variable in dynamical models.

## 3.5 Vertical temperature structure

The mean vertical temperature structure defines the names for the various regions in the atmosphere, as discussed in the first chapter. Here we will consider how this structure arises, along with some basic thermodynamics.

It is a matter of common experience that the atmospheric temperature $T$ declines with height $z$ above the surface. At a certain height, normally around 10 km from the ground at mid-latitudes, this decline ceases and the temperature tends to become isothermal, that is, $dT/dz \to 0$. As also discussed in Chapter 1, the basic reason for this transition from troposphere to stratosphere is the dominance of radiative cooling to space above the tropopause, a topic we shall examine in detail in a later chapter. Here, we use simple physics to calculate approximate expressions for the expected tropospheric temperature gradient and the stratospheric temperature in a simple model where these two quantities, plus the tropopause height (or the surface temperature) completely define the temperature profile as two straight lines (Fig. 3.3).

### 3.5.1 Tropospheric temperature profile

The name 'troposphere' arises because the vertical temperature gradient is maintained mainly by convective overturning, as illustrated schematically in Fig. 3.4. Most of the incoming sunlight heats the ground, and would result in a steep vertical temperature gradient (as shown by the dashed line in the figure) if there were no bulk motions in the atmosphere, so that only conduction and radiation could redistribute the heat. In fact, this 'radiative equilibrium' profile is unstable, because it has cold, dense air overlying warmer, less dense air. Convection takes place and the result

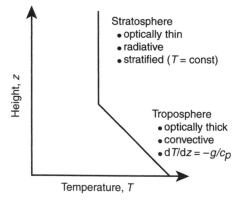

**Fig. 3.3**
A simplified model of the troposphere and stratosphere: the main difference between the two regions is that the troposphere is nearly opaque to infrared radiation and so cools vertically by convection. The stratosphere is nearly transparent and cools by radiation.

**Fig. 3.4**
A schematic of the temperature profiles versus height expected as a result of solar heating, primarily at the surface, if the response of the atmosphere is purely radiative or purely convective. In the troposphere, motions transport heat upwards more efficiently than radiation (or conduction), and so convection results in a less rapid decline in temperature with height.

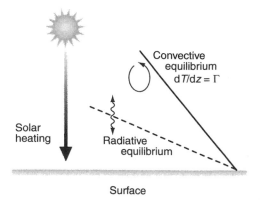

is a less steep profile, the gradient of which we can easily calculate by assuming that the vertical movement of a parcel of air takes place by a small *reversible adiabatic* process (i.e. no exchange of energy or moisture with its surroundings).[36]

From the First Law of Thermodynamics we have for the heat exchange between the parcel and its surroundings during the vertical movement

$$dQ = C_v dT + p dV,$$

where $C_v$ is the specific heat per mole of air. Since conduction is very low and the parcel of air absorbs roughly the same amount of radiation as it emits,[37] $dQ \approx 0$ and this can be treated as an *adiabatic* change.

From the ideal gas law

$$p \, dV + V \, dp = R dT = (C_p - C_v) dT$$

and from the hydrostatic equation

$$V \, dp = -V \rho g \, dz = -Mg \, dz$$

we obtain

$$\frac{dT}{dZ} = -\frac{Mg}{C_p} = -\frac{g}{c_p} = -\Gamma,$$

where $\Gamma$ is called the *adiabatic lapse rate*. The minus sign in this expression arises because $\Gamma$ is defined to be a positive number; so this result predicts that temperature falls as a function of height, as we know from experience it normally does (in the lower atmosphere).

For dry air, the specific heat $c_p = 1.01 \, \text{kJ K}^{-1} \, \text{kg}^{-1}$, giving a value for $\Gamma$ of $9.7 \, \text{K km}^{-1}$. Of course, the real troposphere is rarely dry so the actual value of $c_p$ is larger and $\Gamma$ is smaller. (For pure water vapour the value of $c_p$ is $1.88 \, \text{kJ K}^{-1} \, \text{kg}^{-1}$, which is nearly twice that of dry air). Latent heat effects, discussed later, significantly modify the temperature profile, and in practice, the actual lapse rate is typically around 6 to $7 \, \text{K km}^{-1}$ in the real atmosphere.

Once we have established the value for $\Gamma$ in a particular situation and made a measurement of the actual lapse rate $L$ ($L = -dT/dz$), we can make a statement about whether the atmosphere is stable against convection or not under those conditions. The concept of convective instability can be understood by picturing what happens if a parcel of air is forced to rise from a lower to a higher altitude. If the density of the lifted sample is less than that of the surrounding air, then buoyancy will accelerate it upward, adding to the original displacement; an *unstable* situation. If the density of the displaced parcel is higher than its surroundings, such that it tends to reverse the lifting, the parcel is stable.

[36] The useful concept of a 'parcel' of air invokes an idealized assemblage of molecules, like the contents of a just-buoyant balloon without the skin, that moves a short vertical distance quickly enough that mixing with its surroundings, or transferring heat by conduction or radiation, can all be neglected during the change in pressure and volume that occurs as a result of the motion.

[37] In the troposphere, the atmosphere is 'optically thick', that is, opaque to all but a few 'window' wavelengths (visible light falls in one such window, but they are rare at longer, thermal infrared wavelengths where most of the absorption bands of atmospheric molecules occur, as we shall see later). Thus, radiative exchange by the parcel occurs almost entirely with its immediate surroundings, which are at nearly the same temperature, so there is not much net heating or cooling by this process.

As a parcel of air rises, its change in temperature $dT$ will be

$$dT = -L \, dz$$

whereas for an atmospheric profile that is just at the point of stability

$$dT = -\Gamma \, dz$$

In general the parcel is stable if its density after a small rise $dz$ is greater than or equal to the bulk air density. Since density $\rho$ is inversely proportional to temperature $T$, the parcel needs to end up cooler than the surrounding atmosphere to be stable, that is, $L \, dz$ must be less than $\Gamma \, dz$. In general,

$$\text{if } L < \Gamma \quad \text{the atmosphere is stable}$$

$$\text{if } L > \Gamma \quad \text{the atmosphere is unstable.}$$

In the stable case, a parcel of air that is perturbed (in either direction) will tend to return to its original level. In the unstable case, the perturbed parcel will tend to rise and cool until it achieves the level at which it is stable, that is, at which its density is the same as its surroundings.

### 3.5.2   Stratospheric temperature profile

[38] Leon Phillipe Teisserenc de Bort (1855–1913) was head of the National Meteorological Centre in Paris and later established a private observatory near Versailles where he conducted pioneering experiments with high altitude balloons. He was the first to identify the height (the tropopause) where the lapse rate reaches zero, and to suggest that the atmosphere was divided into two regions, which he named the troposphere and the stratosphere.

The stratosphere was first identified as a distinct region by de Bort[38] at the beginning of the twentieth century, when he noted that balloons tended to stop ascending about 10 km above the surface. On closer investigation, he found that the reason was that the temperature no longer fell rapidly with height, but became isothermal. Convection then ceases, and the atmosphere becomes 'stratified'.

The basic reason for this transition is that, unlike the troposphere below, the stratospheric temperature profile is dominated by radiative rather than dynamical (convective) equilibrium. Under such conditions, we can show that we expect the temperature to vary little with height, (unless there is an additional source of heating, such as that provided by ozone, but we neglect that for the moment).

The stratospheric temperature may be estimated by treating the region as if it were a single slab of gas at temperature $T_S$. This is related to the effective radiative temperature of the Earth, $T_E$ (as defined in Section 2.9), by the expression for the energy balance of the stratosphere, treated as a slab of emittance $e$ (not to be confused with the exponential function, e), as shown in Fig. 3.5 (for a discussion of absorptance and emittance, see Chapter 6). According to Kirchhoff's law (Section 6.2), the absorptance and the emittance are equal, so the slab absorbs a fraction $e$ of the energy flowing upwards from the troposphere below. Since the stratosphere is optically thin overall, this is approximately the same as the energy leaving the planet. The small amount of energy that is absorbed by the stratosphere must be balanced by the emission from it, which takes place equally in both the upward and the downward direction, as shown in Fig. 3.5. According to the Stefan-Boltzmann law (see Section 2.4 and Chapter 6) this is equal to $e\sigma(T_s)^4$ in each direction. Thus we have

$$e\,\sigma(T_E)^4 = 2\,e\,\sigma(T_S)^4$$

from which it follows that

$$T_S = \frac{T_E}{2^{1/4}} = \frac{255}{2^{1/4}} = 215K.$$

Although these simple arguments suggest that the stratospheric temperature remains constant to a great height, in fact the approximate value for $T_s$ just calculated applies to the real atmosphere only for a few km above the tropopause. Above this, temperature is observed to increase, as was shown in Fig. 1.5, up to a maximum value of around 270K near 50 km altitude. This is a consequence of the absorption of solar ultraviolet radiation by the stratospheric ozone layer (Chapter 8).

### 3.5.3 Observed temperature profiles

Figure 3.6 shows some actual, measured temperature profiles made by instrumented weather balloons, known as radiosondes, during various different times of year over Scandinavia. It is immediately apparent that the basic pattern of a constant slope in the troposphere, and a constant value in the stratosphere, is generally present, at least approximately.

The temperature profiles measured by probes to Mars (Fig. 3.7) and Venus (Fig. 3.8) show that the same is true there, and in fact the temperature profiles are simpler than on the Earth, mainly because there are insignificant amounts of ozone in the atmospheres of the other terrestrial planets. On both Mars and Venus, the basic structure with a constant lapse rate troposphere and an isothermal stratosphere predicted by the simple model developed above, can be seen in the observed profiles, although on Mars, in particular, there is a great deal of wavelike structure superimposed. This is due to disturbances with a variety of wavelengths propagating through the

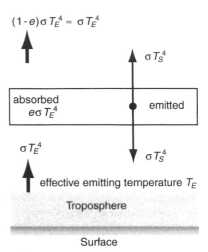

**Fig. 3.5**
An approximate method to calculate the temperature of the stratosphere, by treating it as an optically thin slab of absorptance and emittance $e$, where $e \ll 1$, overlying an opaque troposphere emitting at an equivalent blackbody temperature of $T_E$.

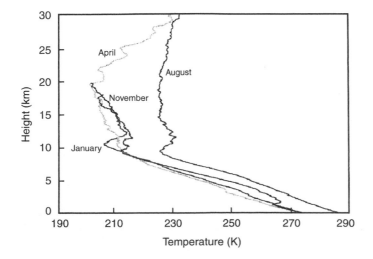

**Fig. 3.6**
Selected vertical temperature profiles measured by individual radiosonde balloon ascents over Scandinavia, in the months indicated, representing the four seasons.

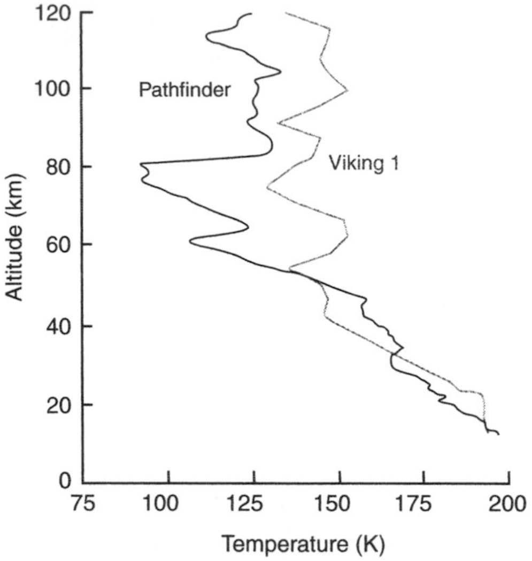

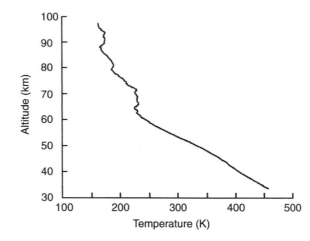

**Fig. 3.7**

Measured temperature profiles for the atmosphere of Mars, measured during the descent of the Viking 1 and Pathfinder landers in 1976 and 1996, respectively.

**Fig. 3.8**

A typical temperature profile for the Venusian atmosphere, derived from measurements of the refraction of the radio signal from the Magellan orbiter as it passed behind the planet on 5 October 1991 at latitude 67N.

[39] The fact that the pressure at the tropopause is two orders of magnitude different, while the mixing ratio of the absorbing gas is about four orders of magnitude different, is also consistent with theoretical expectations, since the opacity in the strong carbon dioxide absorption bands depends on the square root of the absorber amount, as discussed in Section 6.7.

atmosphere at the time of the measurement, a nice example of 'weather' superimposed on the mean structure that we call the climate.

The great depth of the troposphere on Venus reflects the high pressure at the surface (nearly 100 bars). It takes a correspondingly larger vertical distance (about 60 km, compared to about 10 km on the Earth) for the pressure to fall to the point where the overlying atmosphere is thin enough to allow significant cooling by radiation to space.

It also reflects the different compositions of the two atmospheres. The main source of opacity in the stratospheres of both planets is carbon dioxide, but on Earth this is only 0.03% of the total, where on Venus it is nearly 100%. This accounts for the fact that the pressure at the tropopause is about 100 times lower on Venus than on Earth, as we can see by comparing temperature profiles plotted against pressure rather than height (Fig. 3.9).[39] The temperature maximum at about 1 mbar (the stratopause) on Earth stands out particularly in this comparison. If there were no ozone on Earth then the profiles of temperature as a function of pressure on the two planets would be more similar.

## 3.6 The general circulation of the atmosphere

Most of the processes already discussed are linked to the circulation of the atmosphere, which transports heat, momentum, cloud, water, ozone, and pollution around the globe. Of particular importance for the global energy balance is the poleward transfer of energy, to balance the decrease with latitude of the input from the Sun. About half of this transfer, amounting to several petawatts ($1\,\text{PW} = 10^{15}\,\text{W}$) occurs in the atmosphere, and the rest in the oceans. The ocean circulation is coupled to that of the atmosphere

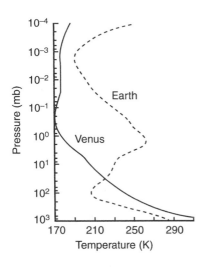

**Fig. 3.9**
Measured temperature profiles in the stratospheres of Venus and Earth, plotted on a common log pressure scale (approximately proportional to height). The observations were made by infrared remote sounding instruments (see chapter 9) on *Pioneer Venus* and *Nimbus* 7, respectively, both in 1979.

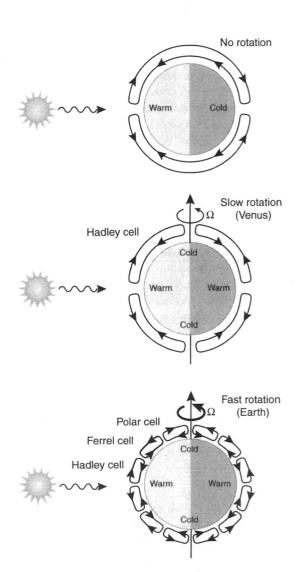

**Fig. 3.10**
A planet with an atmosphere that always kept the same face towards the Sun would have a circulation pattern that flowed from the sub-solar to the anti-solar points (top). Even very slow rotation, as found for example on Venus, changes this so that the axis of symmetry is no longer the line from Sun to planet, but is the rotation axis of the planet (centre). On a fast-rotating planet like the Earth the thermally driven equator-to-pole flow is deflected by the Coriolis acceleration, towards the west for equatorwards motion, and vice versa, in both hemispheres. This renders a single equator-to-pole cell unstable, and instead three cells, known as the Hadley, Ferrel, and Polar cells, are found to dominate the general circulation when wave and chaotic motions are averaged out.

[40] Coriolis and his 'force' are discussed in Section 5.6.1. The rapid rotation of the Earth makes it a key factor affecting the circulation of both the atmosphere and the ocean. Any object moving in a rotating frame of reference experiences an acceleration at right angle to the motion. For example, a bullet fired from a gun at the centre of a rotating table at an object near the edge of the table will miss, because the target moves sideways during the flight of the bullet due to the rotation of the table. The observer on the table who fired the gun sees an apparent force on the bullet diverting it sideways. A similar effect applies to air and water parcels in motion on the rotating Earth.

[41] Edmond Halley in 1686 was the first to explain the prevailing 'trade' winds as part of a large-scale circulation of air that starts with rising air in the tropics, driven by the Sun, which is high in the sky all year round. The rising air is replaced by winds blowing from higher latitudes and balanced by sinking over the colder poles. However, this should result in predominantly north–south winds, whereas the trade winds have a large component from the east. George Hadley (1685–1768) realized that air moving toward the equator travels from regions of relatively low to relatively high eastward surface velocity, and that this would impart a westward motion to the wind. In this thinking he was anticipating the work of Coriolis (q.v.) by 100 yrs. William Ferrel (1817–1891) pioneered the idea that a single cell from equator to pole would not be stable, and that a three-cell profile better fitted the observed wind field.

[42] Suggested by G.W. Paltridge, *Q. J. R. Met. Soc.*, **101**, 475, 1975; see R.M. Goody, *Q. J. R. Met. Soc.*, **126**, 1953, 2000, for a recent discussion.

[43] Boltzmann defined entropy in terms of the number of possible states $W$ that a system can occupy to be $S = k \log(W)$ with constant $k$. When a system takes in a quantity of heat $dQ$ at temperature $T$, the change in entropy is $dS \geq dQ/T$. It remains constant during an adiabatic process, where by definition there is no heat transfer.

by momentum and heat transfer between near-surface winds and the water surface.

If the Earth had a uniform surface and did not rotate, the solar heating would be a maximum at the sub-solar point (the point on the surface where the Sun is directly overhead), and zero everywhere on the night-side. The persistence of this situation would lead to an atmospheric circulation from the sub-solar to the anti-solar point that passed symmetrically around the equator and over the present-day polar regions (Fig. 3.10). We know from the example of Earth's sister planet Venus, which rotates only once every 243 Earth days, that even a very slow rate of spin will redistribute the heating around the equator efficiently enough to make the rotation poles once again tend to be the coldest places on the planet.

We always expect the atmospheric circulation to consist primarily of rising air in the heated region, with descent where cooling exceeds heating, and circulation between the two. For the Venus-like case, the warm air at low latitudes would rise and move polewards, cooling as it went; near the poles, it would descend and move at low levels back towards the equator, completing a large circulation pattern known as a *Hadley cell*. This simple scheme does, in fact, appear to describe the mean atmospheric circulation on Venus, but it is not stable on the more rapidly rotating Earth. Instead, the Coriolis effect[40] adds a component parallel to the equator to the poleward flow in the Hadley cell, which closes the cell, by forcing the north–south flow into an east–west direction, before it can reach high latitudes. Further cells develop and, for reasons that are not intuitively obvious but that can be demonstrated in complex 3-d models, Earth's 24-h rotation causes the development of three circulation cells, the Hadley, Ferrel, and Polar cells, in each hemisphere, rather than one (Fig. 3.10).

This simplified picture[41] is valid only as a long-term mean of the atmospheric general circulation, goes only part-way towards explaining the polewards transport of heat. At least half of the total meridional transport in the atmosphere is, in fact, by eddies, including mid-latitude storm systems and waves and turbulence on a wide range of time and space scales.

A crucial question that we cannot answer with simple physical models or concepts is why the combination of cellular and turbulent transfer leads to the particular equator-to-pole temperature gradient that is observed, although it is a key factor determining the climate and in possible climate change. A guiding principle seems to be[42] that the climate system acts so as to maximize the rate of entropy production. From the definition of entropy $S$,[43] the heat flow $F(\mathrm{W\,m^{-2}})$ from a warm temperature zone at $T_1$ to a cooler one at $T_2$ is related to the entropy production rate $dS/dt$ by

$$\frac{dS}{dt} = \frac{F}{T_2} - \frac{F}{T_1}.$$

If this were considered a model for the heat flux polewards in the Hadley circulation for example, then if all motion were stopped, $F$ would be zero and so would entropy production. In the other extreme, if the motions were very rapid then $T_2$ would tend to the same value as $T_1$ and $dS/dt$ would again be zero. The actual situation is somewhere in between and is postulated, with some support from studies using general circulation models, to be that for which $dS/dT$ finds a maximum value.

Thus it appears that dynamical energy transfer within the climate system not only tends towards a state of maximum entropy, as required by the second law of themodynamics, but also that it moves in this direction at the fastest possible rate. The explanation, broadly speaking, is that the processes involved are complex and chaotic, and so offer an essentially infinite number of multiple pathways towards the state of maximum disorder. (The absorption, transfer, and emission of radiation by the climate system, which involves a larger rate of entropy production than the dynamical transport of heat, does not have this property and does not behave in this way.) Processes involving turbulence are difficult to represent accurately in numerical models, and it has been proposed that checking that the dynamical schemes in climate forecasting models satisfy the principle of maximum entropy production is a good test of their validity.

## Further reading

*The Physics of Atmospheres*, J.T. Houghton, 3rd edition, Cambridge University Press, Cambridge, 2002.

*An Introduction to Atmospheric Physics*, D.G. Andrews, Cambridge University Press, Cambridge, 2000.

*Fundamentals of Atmospheric Physics*, M.L. Salby, Academic Press, San Diego, 1996.

## Questions

1. Why does the temperature of the lower atmosphere decrease with height? Define the term *lapse rate* and estimate its value on (a) Earth, (b) Mars, (c) Venus, stating any assumptions that you make. In the case of the Earth, consider both dry and saturated atmospheres separately.
2. Estimate the total mass (in tonnes) of (a) carbon dioxide (b) ozone in the global atmosphere, stating any assumptions made.
3. Why does pressure fall with height in the atmosphere? Derive an expression for pressure as a function of height in an isothermal atmosphere. Introduce the concept of a scale height, and explain what this means. Estimate the height above the surface of the level at which half of the mass of the atmosphere lies above and the other half below.
4. A parcel of air is lifted slowly from the ground, where the temperature is 295K, to an elevation of 5 km, and then returned rapidly to the ground. Estimate the air parcel temperature at altitude and after it returns to the ground, explaining any assumptions made.
5. The temperature in the lower stratosphere is approximately constant as a function of height because the overlying atmosphere is optically thin. Estimate the temperature of the Earth's stratosphere given that the effective temperature at which the Earth radiates to space is 255K.
6. The pressure at the base of the stratosphere (the tropopause) is expected to depend approximately on the square root of the amount of overlying carbon dioxide, because the strong absorption bands of $CO_2$ are optically thick. Show quantitatively that this is the case by comparing figures for Earth and Venus.
7. Explain briefly the resolution of the apparent paradox in the two preceding questions, which occurs between the statements about optical thickness in the stratosphere.

# 4

# Clouds and aerosols

## 4.1 Introduction

Clouds are, in many ways, the most crucial part of the climate system, because they have a large effect on the energy balance of the atmosphere and surface, but are very variable and difficult to predict. On a warming Earth, would the amount of cloud cover increase? Would this result in an increase in the albedo of the planet, and would the cooling effect of this dominate the warming effect of adding further infrared opacity to the greenhouse effect, or vice versa? The answers to even these basic questions remain elusive. We will look at the role of clouds in the greenhouse effect and at attempts to model their contribution, in later chapters. Here we look at the basic physics of cloud formation and growth.

## 4.2 Potential temperature and entropy

To understand cloud formation we return to the basic question of whether a particular one-dimensional representation of the atmospheric temperature profile is stable against vertical convection or not, and examine the role of water vapour in determining stability. Potential temperature and entropy are thermodynamic quantities derived from composition, temperature, and pressure that turn out to be useful for this purpose.

The vertical temperature profile in the atmosphere tends towards radiative equilibrium between the heating caused by the absorption of incoming sunlight, and the cooling to space by long-wavelength infrared emission, at every level. However, for any reasonable atmospheric composition, a profile based purely on radiative equilibrium is always unstable in the troposphere, resulting in vertical motions (convection) that generate many of features of the weather at the surface.

Convection occurs even on the night side of the planet, because it is a relatively fast process, with characteristic time scales of a few hours. Perturbations away from radiative equilibrium, on the other hand, relax back to equilibrium on time scales of more than a month in the troposphere. Therefore, we expect the observed temperature profile to be much closer to convective than to radiative equilibrium at all times.

To compute the degree of instability corresponding to a given $T(p)$ profile it is sometimes convenient to work with the *specific volume* $\alpha$ (the inverse of the density, so $\alpha = 1/\rho$) of the adiabatically displaced parcel and compare it to the specific volume of the upper layer to which it is moved. If the parcel has a greater specific volume, its density will be less and the displacement will thus be unstable, leading to further up ward motion. If the specific volume decreases with increasing height, a restoring force tends to resist the vertical motions leading to stability.

Another useful definition is the *potential temperature* $\theta$, the temperature that a parcel of air at that point would achieve if it were moved adiabatically to a standard pressure, usually 1 bar (i.e. the surface). For an adiabatic change we have, from the first law of thermodynamics and the ideal gas equation (cf. Section 3.5)

$$C_p \, dT = V \, dp = \frac{RT}{p} \, dp$$

so

$$\frac{C_p}{R} \frac{dT}{T} = \frac{dp}{p},$$

which can be integrated to give:

$$\frac{C_p}{R} \log T = \log p + \text{const.}$$

If $T$ is the initial temperature and $\theta$ the final temperature (at $p = p_0$, so $\theta$ is the potential temperature by definition) we obtain:

$$\theta = T \left( \frac{p_0}{p} \right)^{R/C_p}$$

called *Poisson's equation*,[44] where the value of the exponent $R/C_p$ is about 0.28. To find the variation of potential temperature with height we differentiate to get

$$\frac{d\theta}{dz} = \theta \left( \frac{1}{T} \frac{dT}{dz} - \left( \frac{R}{C_p} \right) \frac{1}{p} \frac{dp}{dz} \right)$$

and using the hydrostatic equation $dp = -\rho g \, dz$, with $pV = RT$ and $c_p = C_p / V$

$$\frac{d\theta}{dz} = \frac{\theta}{T} \left( \frac{dT}{dz} + \frac{g}{c_p} \right)$$

so

$$\frac{d\theta}{dz} = \frac{\theta}{T} \left( \frac{dT}{dz} + \Gamma \right),$$

where the expression inside the parentheses is the adiabatic lapse rate $\Gamma$, a positive number, plus the actual lapse rate $dT/dz$, which is negative (see Section 3.5). Thus, the atmosphere is stable against convection if $d\theta/dz > 0$ and unstable if $d\theta/dz < 0$.

[44] Simeon-Denis Poisson (1781–1840) was a protegé of Laplace and Lagrange, who in 1806 replaced Fourier as Professor of Mathematics at the Ecole Polytechnique in Paris.

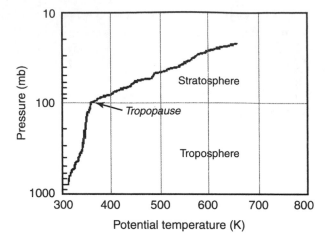

**Fig. 4.1**
Potential temperature as a function of log
pressure (which is approximately
proportional to height) as derived from
balloon measurements at the same time
and location as Fig. 3.1.

Figure 4.1 shows a vertical profile of potential temperature, derived from balloon measurements, from which it can be seen that the atmosphere, on this occasion, was just stable in the upper troposphere, and very stable in the lower stratosphere, with a sharp transition between the two.

Potential temperature is related to the entropy profile of the atmosphere, as we can see if we write the first law of thermodynamics for 1 mole in the form:

$$\mathrm{d}Q = C_p\,\mathrm{d}T - V\,\mathrm{d}p$$

and introduce the entropy, $\mathrm{d}S = \mathrm{d}Q/T$, whence

$$\mathrm{d}S = C_p\frac{\mathrm{d}T}{T} - \frac{V}{T}\,\mathrm{d}p$$

and

$$\frac{V}{T} = \frac{R}{p}.$$

Hence

$$S = C_p\log T - R\log p + \text{const.}$$

From the definition of potential temperature,

$$S = C_p\log\theta - R\log\,p_0 + \text{const}$$

or, with the constants gathered together,

$$S = C_p\log\theta + \text{const.}$$

Considering changes in entropy we have

$$\Delta S = C_p\Delta\log\theta$$

from which we see that $\log\theta$ is essentially a scale of entropy.

For a forcibly uplifted air parcel,

$$\frac{\mathrm{d}S}{\mathrm{d}z} = \frac{C_p}{\theta} \frac{\mathrm{d}\theta}{\mathrm{d}z}$$

$$= \frac{C_p}{T}\left(\frac{\mathrm{d}T}{\mathrm{d}z} + \Gamma\right).$$

Since the adiabatic lapse rate $\Gamma$ is always positive and $\mathrm{d}T/\mathrm{d}z$ negative, this result is telling us that the atmosphere will be convectively unstable ($\Gamma$ numerically less than $\mathrm{d}T/\mathrm{d}z$) whenever entropy decreases upward along the gravitational vector, and vice versa. The measured profile in Fig. 4.1 shows potential temperature and hence entropy increasing upwards slightly in the troposphere, and sharply in the stratosphere. This is consistent with stability everywhere, but marginal in the troposphere and large in the stratosphere.

## 4.3 Potential energy and available potential energy

Consider a column of air of unit cross-sectional area. The total energy of an element of mass $\mathrm{d}m$ at height $z$ and pressure $p$ is

$$\mathrm{d}E = \mathrm{d}E_i + \mathrm{d}E_p,$$

where $E_i$ is internal energy and $E_p$ is potential energy.

$$\mathrm{d}E = \frac{C_v}{M}T\,\mathrm{d}m + zg\,\mathrm{d}m$$

from the hydrostatic equation

$$\mathrm{d}m = \rho\,\mathrm{d}z = -\frac{\mathrm{d}p}{g}$$

integrating from the ground at pressure $p = p_0$ to the top of the atmosphere ($p = 0$):

$$E_i = \int_{p_0}^{0} \frac{C_v T}{Mg}\,\mathrm{d}p$$

and

$$E_p = -\int_{p_0}^{0} \frac{RT}{mg}\,\mathrm{d}p$$

hence

$$\frac{E_p}{E_i} = \frac{R}{C_v} = \gamma - 1 = \text{const.}$$

$$E = -\int_{p_0}^{0} \frac{C_p T}{Mg}\,\mathrm{d}p = -\int_{p_0}^{0} \frac{1}{\Gamma}T\,\mathrm{d}p.$$

Assuming a mean value for $T$ of 250K, $E$ is about $2.5 \times 10^9$ $Jm^{-2}$. The area of the Earth is about $5.1 \times 10^{14}$ $m^2$ giving $E \sim 1.3 \times 10^{24}$ J for the total potential energy of the atmosphere. This is a large number, but it is intuitively obvious that most of it cannot be available to do work.[45]

A more useful concept is the *available potential energy* present as a function of height in a column of air. This is calculated using the difference values of temperature, humidity, and pressure, that is, their actual values minus some standard profile. The available potential energy then provides a measure of the kinetic energy that the whole atmospheric column could acquire, starting from rest and moving vertically due to convection.

## 4.4  Humidity

The origin of water vapour in the atmosphere is surface evaporation, especially from the ocean, and of course this is recycled, usually at some other location than its origin, as rain or other precipitation. The concentration of water vapour in the air varies with latitude and height from almost zero to several % of the mass of a vertical column. There are several different ways to express the value of the humidity, the most basic of which are the mass and volume mixing ratios.

The molecular weight (molar mass) of a sample of air will depend on the amount of water vapour it contains. Consider the hydrostatic equation for a mixture of gases:

$$dp = dp_1 + dp_2 = -\rho_1 g \, dz - \rho_2 g \, dz$$

$$= -\left(\frac{M_1 p_1}{RT} + \frac{M_2 p_2}{RT}\right) g \, dz = -\frac{M' p}{RT} g \, dz = -\rho \, dz,$$

where

$$M' = \frac{M_1 p_1 + M_2 p_2}{p}$$

is the *effective molecular weight* $M'$ of the mixture, a pressure (number) weighted mean of each constituent.

For dry air

$$M_{dry} = \frac{78.09 M_{N_2} + 20.95 M_{O_2} + 0.93 M_{Ar}}{99.97}$$

inserting the values for the molecular weights of nitrogen, oxygen, and argon (28.02, 32, and 39.94 $kg\,kmol^{-1}$, respectively) we obtain

$$M_{dry} = 28.96 \text{ kg kmol}^{-1}.$$

For moist air

$$M_{wet} = \frac{M_{dry} p_{dry} + M_{H_2O} p_{H_2O}}{p},$$

where $p_{H_2O}$ is the water vapour partial pressure and $p$ the pressure of the damp air.

[45] On the Earth, that is. On a gas giant planet like Jupiter, which consists mostly of atmosphere, internal energy comparable to the sunfall is released from the interior of the planet as a result of a slow contraction in its size. As a result, the observable, outer part of Jupiter's atmosphere is heated from above and below in approximately equal proportions.

The *mass mixing ratio* $x_m$ is the ratio of the mass of water vapour to the mass of dry air in a volume $V$ with total pressure $p$. Dalton's law of partial pressures states that the total pressure $p$ is the sum of the pressure that would be exerted by each gas if it were present on its own, so

$$p = p_a + p_{H_2O},$$

where $p_{H_2O}$ and $p_a$ are the *partial pressures* of water vapour and dry air respectively.

If $\rho_{H_2O}$ and $\rho_a$ are the corresponding densities, using $\rho = Mp/RT$ and noting that $M_{H_2O}/M_a = 18.015/28.964 = 0.622$ we obtain

$$x_m = \frac{\rho_{H_2O}}{\rho_a} = \frac{0.662\, p_{H_2O}}{p_a}.$$

The *volume mixing ratio* $x_v$ is the ratio of the volume of water vapour to that of dry air if the constituents were separated at pressure $p$. It follows that

$$x_v = \frac{p_{H_2O}}{p_a} = \frac{n_{H_2O}}{n_a},$$

where $n$ is the number of molecules of each type in the volume $V$.

The amount of water in the atmosphere is often quoted in terms of *relative humidity, dew point*, or *virtual temperature*, since these are often easier to relate to than absolute humidity or mixing ratio so far as the human environment is concerned. If these are encountered, it will be necessary to know how they are defined and how to relate them to more basic atmospheric variables.

The relative humidity $U$ is a dimensionless number that relates the amount of water in the air to the equilibrium value (at saturation). If SVP is the saturated vapour pressure at temperature $T$, then

$$U = \frac{p_{H_2O}}{SVP}$$

and the mass mixing ratio is given, in terms of relative humidity, by

$$x_m = \frac{0.622(U \times SVP)}{p - (U \times SVP)}.$$

The dew (or frost) point $T_D$ is the temperature to which air has to be cooled for condensation (or sublimation) to occur (see Fig. 4.2). It is the temperature where the saturated vapour pressure is equal to the partial pressure $p_{H_2O}$.

The virtual temperature, $T_{virtual}$, of moist air is defined as the temperature of dry air that has the same density. Since the molecular weight of water vapour is less than that of dry air, $T_{virtual}$ is always greater than actual

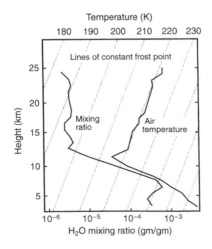

**Fig. 4.2**
Profiles of air temperature and water vapour mass mixing ratio, measured by balloon-borne instruments at 52°N.

temperature $T$. Since

$$\rho_{\text{wet}} = \frac{M_{\text{wet}}p}{RT} = \frac{M_{\text{dry}}p}{RT_{\text{virtual}}},$$

we have

$$T_{\text{virtual}} = T\frac{M_{\text{dry}}}{M_{\text{wet}}} = \frac{T}{\left\{1 - (p_{\text{H}_2\text{O}}/p)\left(1 - (M_{\text{H}_2\text{O}}/M_{\text{dry}})\right)\right\}}.$$

Figure 4.2 shows some typical measurements of temperature and water vapour mixing ratio, both of which fall with increasing height in the upper part of the troposphere, and then become nearly constant in the lower stratosphere. The slight rise in stratospheric temperature, following a line of constant frost point, suggests that the air is saturated at the very low temperatures found just above the tropopause, even though the mixing ratio is only about 4 ppm by volume.

## 4.5  Thermodynamics of moist air

The conditions for the stability of moist air are the same as for dry air, provided that we allow for the difference in molecular weight and heat capacity, and assume that saturation does not occur. However, once the rising air cools to the point where it becomes saturated, further rising will result in condensation and the latent heat released will reduce the cooling effect and change the lapse rate. To calculate the new value, we need to include an extra term in the First Law expression thus:

$$dQ = C_{\text{v}}\,dT + p\,dV + \frac{L}{M_{\text{H}_2\text{O}}}\,dm = 0,$$

where $m$ is the mass of water in 1 mole of air and $L$ is the molar latent heat of water. From the gas law we also have

$$p\,dV + V\,dp = C_{\text{p}}\,dT - C_{\text{v}}\,dT$$

Combining these equations and differentiating with respect to height $z$ we obtain

$$C_{\text{p}}\frac{dT}{dz} - V\frac{dp}{dz} + \frac{L}{M_{\text{H}_2\text{O}}}\frac{dm}{dz} = 0$$

we can relate $m$ to the partial pressure $p_{\text{H}_2\text{O}}$ of water by

$$m = \frac{M_{\text{H}_2\text{O}}p_{\text{H}_2\text{O}}}{RT}V = \frac{M_{\text{H}_2\text{O}}p_{\text{H}_2\text{O}}}{p}.$$

Hence,

$$\frac{L}{M_{\text{H}_2\text{O}}}\frac{dm}{dz} = L\left(\frac{1}{p}\frac{dp_{\text{H}_2\text{O}}}{dT}\frac{dT}{dz} - \frac{p_{\text{H}_2\text{O}}}{p^2}\frac{dp}{dz}\right).$$

Now we use the Clausius-Clapeyron equation,[46] which describes the vapour pressure as a function of temperature:

$$\frac{\mathrm{d}p_{H_2O}}{\mathrm{d}T} = \frac{L}{T(V_v - V_l)} = \frac{Lp_{H_2O}}{RT^2}$$

where $V_v$ and $V_l$ are the molar volumes of water vapour and liquid respectively. Since $V_v \gg V_l$ and $p_{H_2O}V_v = RT$, we have

$$\frac{\mathrm{d}T}{\mathrm{d}z}\left(C_p + \frac{L}{p}\frac{Lp_{H_2O}}{RT^2}\right) = \frac{\mathrm{d}p}{\mathrm{d}z}\left(v + \frac{Lp_{H_2O}}{p^2}\right)$$

and using the hydrostatic equation $\mathrm{d}p/\mathrm{d}z = -\rho g$, we finally obtain the expression for the *moist adiabatic lapse rate*:

$$-\frac{\mathrm{d}T}{\mathrm{d}z} = \Gamma_{sat} = \frac{\Gamma_{dry}\left(1 + (Lp_{H_2O}/pRT)\right)}{\left(1 + (L^2 p_{H_2O}/C_p pRT^2)\right)}.$$

The saturated lapse rate $\Gamma_{sat}$ is always less than the dry value $\Gamma_{dry}$, although at low temperatures they are nearly the same. At higher atmospheric temperatures $\Gamma_{sat}$ can fall as low as $0.35\Gamma_{dry}$. Moist air is absolutely stable if the absolute value (i.e. ignoring the minus sign) of $\mathrm{d}T/\mathrm{d}z$ is less than $\Gamma_{sat}$. If $\Gamma_{sat} < |\mathrm{d}T/\mathrm{d}z| < \Gamma_{dry}$, the atmosphere is said to be *conditionally stable*. It remains in this state unless forced above the condensation level, when it will become unstable and continue to rise, generally leading to the formation of clouds.

We can now distinguish between vertical temperature profiles for three cases (Fig. 4.3). The first is purely radiative equilibrium, representing the hypothetical situation in which there is no vertical movement of the air, so the profile is determined by balancing the heating caused by the absorption of incoming sunlight, the cooling to space by long-wavelength infrared emission, and the radiative exchange between atmospheric layers at different heights. The next concept is an atmosphere with no water vapour, but where the vertical transport of heat by bulk movement of air (convection) is allowed. This is a closer approximation to reality, because convection is more efficient than radiation at moving heat vertically when the atmosphere is optically thick (i.e. opaque to radiation, at all or most wavelengths; see Chapter 6). Finally, the most realistic case is the one just calculated where significant amounts of water vapour are present and latent heat transfer by water vapour is an important factor in determining the lapse rate in the troposphere. The large difference between the three cases emphasizes the importance of dynamics and latent heat for the climate at the surface. The fact that the profiles converge above the tropopause shows also that these processes are much less important in the stratosphere, where the atmosphere is optically thin (i.e. transparent at most wavelengths) and radiation dominates.

[46] Emile Clapeyron (1799–1864) was an engineer interested in steam locomotives. He developed a formula for the heat of vaporization of a liquid, which was later extended by Rudolph Clausius (1822–1888) to show that the vapour pressure of a liquid is not a linear function of temperature, but increases exponentially as the boiling point of the liquid is approached.

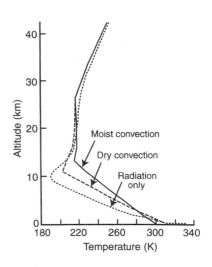

**Fig. 4.3**
Representative vertical temperature profiles calculated assuming (top curve) convective equilibrium in a moist atmosphere, (middle) convective equilibrium in a dry atmosphere, and (bottom) radiative equilibrium. See: Thermal equilibrium of the atmosphere with a given distribution of relative humidity, S. Manabe and R.T. Wetherald, *J. Atmos. Sci.*, **24**, 241–259, 1967.

## 4.6 Condensation processes and cloud formation

The saturated vapour pressure of water is a strong function of temperature over both water and ice. When considering cloud droplet formation and growth, the SVP of interest is normally that at the surface of a more-or-less spherical droplet; then SVP also depends on the curvature of the droplet, and on the nature and concentration of any impurities that may be dissolved in the droplet. In addition, SVP is different for particles of ice and super-cooled water at the same temperature, the former being substantially lower. The combination of all of these factors determines the growth rate of very small droplets in the atmosphere, leading to the formation of clouds of much larger droplets, and eventually to the production of rain. Clouds are import-ant, not just because of their role in the hydrological cycle, but because they absorb and reflect solar and terrestrial radiation, so they have a large part to play in the radiative transfer of energy within the climate system.

The saturated vapour pressure above a curved surface, like that of a cloud droplet, is not the same as that above a flat surface of the same liquid, but is higher because molecules can evaporate more easily from a curved surface. Thus, if we imagine a droplet suspended in the air above an open surface of water like a lake, both assumed to be pure $H_2O$, the ambient vapour pressure will be that pertaining to the flat surface while that near the droplet will be higher, and the droplet will tend to shrink by diffusion of water vapour outwards, down the humidity gradient.

To look at this qualitatively we use *Kelvin's formula*[47] for the vapour pressure $p_v(r)$ over a convex surface of radius $r$. This can be written in the form

$$p_v(r) = p_v(\infty) \exp\left(\frac{2\sigma}{r}\frac{M}{\rho_l RT}\right),$$

where $p_v(\infty)$ is the equivalent value over a plane surface $\sigma$ and $M$ are the surface tension and molecular weight of the liquid and $\rho_l$ the density of the liquid. For water, $\sigma \sim 8 \times 10^{-2}\,\mathrm{N\,m^{-1}}$. The *saturation ratio S*, defined as the vapour pressure relative to that above a flat surface for the same liquid and temperature, is defined by

$$S = \frac{p_v(r)}{p_v(\infty)} = \exp\left(\frac{a}{r}\right),$$

where $a$ is a constant in the same units as $r$, usually micrometres ($\mu$m), and temperature $T$ is taken to be constant in the vicinity of the drop. The *supersaturation S'* (in %) is defined as $(S - 1) \times 100$, and inserting values we find that $S' \approx 10^{-3}\,r^{-1}$ with $r$ in $\mu$m. Thus, $S'$ is 0.1% for $r = 1\,\mu$m and 10% for $r = 10$ nm ($10^{-8}$ m).

The air is always supersaturated above a small drop so it will tend to evaporate even if the surrounding air is saturated. In order for the droplet to grow and become part of a cloud, it must either be in air that is itself super-saturated to an even greater degree, or it must use some other mechanism to grow until its SVP falls below that of the surrounding air. The critical radius $r^*$, at which this occurs, is found by rearranging Kelvin's formula

---

[47] William Thompson, Lord Kelvin (1824–1907) was seeking to explain what happened to the air pressure above a menis-cus in a capillary tube, when he developed a formula that also usefully applies to droplets suspended in the atmosphere. It can be derived as follows.

Consider the formation of a droplet of radius $r$ by condensation from vapour, where the free energy per molecule is $\phi_v$ in the vapour and $\phi_l$ in the liquid. If $n$ is the number of molecules in the drop, then the total change in free energy as a result of the formation of the drop is

$$\Delta G = (\phi_l - \phi_v)n + 4\pi r^2\sigma,$$

where $\sigma$ is the surface tension of the liquid. If the vapour pressure changes from $p_v$ to $p_v + dp_v$, then

$$d\phi_v = v_v dp_v,$$

$$d\phi_l = v_l dp_v,$$

where $v_v$ and $v_l$ are the volumes occupied by one molecule in the vapour and the liquid, respectively. Since $v_v \ll v_l$,

$$d\phi_l - d\phi_v = V_l\,dp_v = -\frac{kT}{p_v}\,dp_v,$$

treating the vapour as an ideal gas at tem-perature $T$ where $k$ is Boltzmann's constant. Integrating both sides gives

$$\phi_l - \phi_v = kT \ln\left[\frac{p_v(r)}{p_v(\infty)}\right]$$

whence

$$\Delta G = nkT \ln\left[\frac{p_v(r)}{p_v(\infty)}\right] + 4\pi r^2\sigma.$$

The droplet is stable when $d(\Delta G)/dr = 0$, that is, when

$$\ln\left[\frac{p_v(r)}{p_v(\infty)}\right] = \frac{2\sigma}{r\rho_l}\frac{M}{RT},$$

where $M$ is the molar mass of the liquid, $\rho_l$ its density, and $R$ is the molar gas constant.

to obtain

$$r^* = \frac{2\sigma M}{\rho_l RT \log[S]}.$$

The probability of a droplet reaching a stable size by homogeneous nucleation (i.e. the random congregation of water molecules) is very small under most conditions. Consider, for example, that to make a droplet with a radius of 0.5 nm requires the simultaneous collision of about 17 molecules, and around 750% supersaturation of the ambient air.

It is the presence of tiny soluble particles in the atmosphere to act as *condensation nuclei* that makes the formation of large droplets possible. Only soluble (hygroscopic) particles make effective condensation nuclei—soot, dust, etc. are little more effective than homogeneous nucleation. A common source of soluble nuclei for cloud formation is the evaporation of sea spray, which forms large numbers of tiny salt crystals small enough to remain airborne essentially indefinitely. The crystals accumulate water molecules until a solution results; the effect of the dissolved solvent is to dramatically reduce the SVP adjacent to the surface of the droplet and thus induce further growth.

Only a very small quantity of salt is required; around $10^7$ molecules or about $10^{-15}$ g of NaCl is enough to start the process. Such a particle has a radius of only about 0.05 μm, much smaller than the wavelength of visible light. An electrical charge can have a similar effect since it also reduces the surface tension of the droplet, however this effect is normally much less important in the atmosphere. Aerosol particles that exist in the atmosphere both naturally, for example, from volcanoes, and as a result of air pollution, are also an important and growing source of cloud nuclei. Some scientists believe that anthropogenic aerosols, mostly from fossil fuels, which release an estimated $10^{14}$ g of sulphur into the atmosphere each year, already dominate cloud seeding worldwide. If this is the case, then this is another way in which human activities can have a major effect on the global climate.

The presence of a solute lowers the saturation vapour pressure according to *Raoult's Law*,[48] which states simply that the SVP of a drop of radius $r$ depends directly on the concentration of solute, $n_0/(n_0 + n)$, where $n$ and $n_0$ are the numbers of molecules of solute and of water respectively. Thus:

$$S(r) \sim \frac{n_0}{n_0 + n} \sim \left(1 + i\frac{n}{n_0}\right)^{-1} = \left(1 + \frac{b}{r^3}\right)^{-1}.$$

[48] Francois Raoult (1830–1901) proposed in 1886 that the partial pressure of the vapour in equilibrium with a solution is proportional to the ratio of solvent to solute molecules.

The introduction of the constant $i$ (which varies with solute but typically has a value ~2, and is known as van't Hoff's factor)[49] improves the approximation by allowing for the degree of ionic dissociation in solvent and solute. The number of solute molecules, $n$, is a constant for a droplet formed on a given nucleus, while for a spherical droplet the number of water molecules $n_0$ is proportional to $r^3$. The constants are gathered into a single constant $b$, which has dimensions of $(\mu m)^3$.

[49] Jacobus van't Hoff (1852–1911) showed in 1885 that the thermodynamic laws of gases also apply to dilute solutions with the addition of a constant factor $i$ to allow for the ionization that occurs in solutions.

Clouds and aerosols

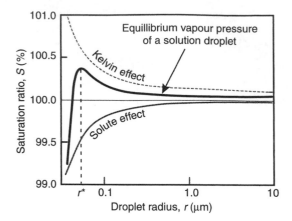

**Fig. 4.4**
The Kohler curve (thick line) for the vapour pressure at the surface of a droplet containing solute, which reduces the vapour pressure, as a function of droplet radius, on which the vapour pressure depends inversely through the Kelvin effect. The net effect is a saturation ratio (SVP relative to a flat surface) with a strong peak at the critical radius $r^*$ for droplet growth. Once the droplet is larger than this, adding water reduces the SVP at its surface and encourages further growth, and vice versa.

Now the SVP over a droplet is the product of two terms:

$$S(r) = \exp\left(\frac{a}{r}\right)\left(1 + \frac{b}{r^3}\right)^{-1}.$$

Expanding each term as a series and neglecting higher-order terms we obtain:

$$S(r) \approx 1 + \frac{a}{r} - \frac{b}{r^3}.$$

This approximation is reasonably valid for values of $a$ and $b$ that apply in the real atmosphere, and illustrates clearly how the 'curvature' or Kelvin term, which tends to increase the saturation at the droplet surface, competes with the 'solute' or Raoult term tending to decrease it.

A plot of this function is called a *Kohler curve*[50] (Fig. 4.4) and it has a maximum at the critical values of $r$ and $S$ above which the particle will grow, and below which it will shrink, in saturated (but not supersaturated) air. By differentiating we find that these are

$$r^* = \sqrt{\frac{3b}{a}} \qquad S^* = 1 + \sqrt{\frac{4a^3}{27b}}.$$

Typical values of $a$ and $b$ correspond to sub-micron sizes of critical radius and a few percent supersaturation (see Question 8).

## 4.7 Growth of cloud droplets

Assuming that condensation nuclei are available and that the appropriate degree of supersaturation exists, a cloud droplet will grow by diffusion of

[50] Introduced by H. Kohler in his paper, The nucleus in and the growth of hygroscopic droplets, *Trans. Faraday Soc.*, 32, 1152–1161, 1936.

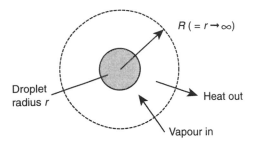

vapour towards the droplet followed by condensation. As this occurs, latent heat is released which tends to increase the temperature of the droplet. Some of this heat is conducted away, but overall there is an increase in the temperature of the droplet that tends to reduce the condensation rate.

The growth of a drop of mass $m$ and radius $r$ is given by the diffusion law

$$\frac{dm}{dt} = 4\pi R^2 D \frac{d\rho}{dR},$$

where $D$ is the diffusion coefficient of water vapour in air, and $\rho$ is the vapour density.

Now we integrate the equation over the radial distance $R$, from the surface of the droplet at $R = r$ to $R = \infty$ (Fig. 4.5) to get:

$$\frac{dm}{dt} = 4\pi r D (\rho_\infty - \rho_r).$$

The vapour densities $\rho_r$ and $\rho_\infty$ are the boundary conditions at, and far from, the surface of the drop. We also have

$$\frac{dm}{dt} = 4\pi r^2 \frac{dr}{dt} \rho_l,$$

where $\rho_l$ is the density of liquid water, so, equating these two expressions for $dm/dt$ we get

$$\frac{dr}{dt} = \frac{D}{r} \left( \frac{\rho_\infty - \rho_r}{\rho_l} \right)$$

for the rate of growth of the droplet due to the diffusion of vapour if its temperature, and hence $\rho_r$, remains constant.

However, the drop will gain heat $Q$ due to release of latent heat $L$ at a rate given by

$$\frac{dQ}{dt} = L \frac{dm}{dt} = 4\pi r L D (\rho_\infty - \rho_r).$$

There is also a heat loss due to the thermal conductivity of air, $\lambda$. The heat flow through the surface of the drop at radius $r$ and its surroundings at

**Table 4.1** The growth history of a droplet with initial radius of 0.75 μm, containing, $10^{-14}$ g of NaCl at an atmospheric level where the relative humidity is 100.05%, pressure is 900 mb, and $T = 273$K

| Radius (μm) | Time |
|-------------|---------|
| 0.75 | 0 |
| 1 | 24 s |
| 2 | 130 s |
| 4 | 1,000 s |
| 10 | 2,700 s |
| 20 | 2.4 h |
| 30 | 4.9 h |
| 40 | 12.4 h |

distance $R$ is

$$\frac{dQ}{dt} = 4\pi r^2 \lambda \frac{dT}{dR}$$

again integrating from $R = r$ to $R = \infty$:

$$\frac{dQ}{dt} = 4\pi r \lambda (T_\infty - T_r).$$

Equating the two expressions for $dQ/dt$, we find that the droplet is warmer than its surroundings by an amount

$$(T_r - T_\infty) = \frac{LD(\rho_r - \rho_\infty)}{\lambda}.$$

These equations can be solved, using the Clausius-Clapeyron and Kohler expressions, to obtain the growth rate of a drop under a given set of conditions. This requires rather tedious numerical integration and introduces no new physics, so we will not attempt it but instead show some typical results in Table 4.1.

Under most atmospheric conditions, the growth rate of a droplet by diffusion of vapour is very slow: a droplet may take several hours to grow to a radius of a few tens of micrometres. There are other processes that are faster, and which also commonly occur in the real atmosphere: growth by freezing, and growth by coagulation.

## 4.8  Growth of ice crystals

Since the SVP over ice is less than that over supercooled water, an ice crystal will grow more rapidly than a water drop at the same temperature. Moreover, ice crystals will grow at the expense of water droplets if both are present together. Under the right conditions, ice crystals can grow to 100 μm and more in only a few minutes. It is very common for precipitation to originate in this way, including rain, since in all but very cold conditions the ice particles melt before reaching the ground.

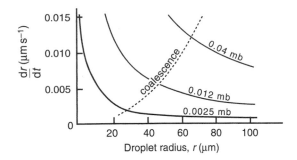

**Fig. 4.6**
Examples of the rate of growth of cloud droplets in a cloud with a fixed liquid water content of
1 g cm$^{-3}$ by coalescence (dashed line) and by condensation (solid lines), assuming increasingly
supersaturated water vapour pressures (shown in mb above the saturation value) in the air
surrounding the droplet. A small droplet will grow primarily by condensation until it reaches
the size where coalescence is faster; after which the latter takes over and proceeds rapidly.

## 4.9  Collision and coalescence

Once particles begin to fall, larger particles will collide with the smaller
ones and grow by coalescence. If the cloud depth is adequate, large drops
are produced in this way. Whereas growth by condensation tends to slow
down as particles get larger, large particles formed by coalescence grow
ever faster, as Fig. 4.6 shows. This growth rate will be further enhanced by
the updraughts that are often found in clouds.

## 4.10  Aerosols

Aerosols are tiny particles suspended in the air, generally much smaller than
the particles that make up visible clouds. Aerosols may be invisible to the
eye, or at most present the appearance of a fine haze or 'turbidity,' and still
be very important within the climate system, primarily as

(1) nuclei for cloud droplet formation and growth,
(2) scatterers of incoming sunlight, increasing the albedo of the planet,
(3) absorbers of both solar and long-wave radiation.

Some aerosols are of natural origin, from volcanoes, dust storms, forest and
grassland fires, pollen, and sea spray. Human activities such as the burning
of fossil fuels, and agricultural or industrial processes, also generate aerosols
(Fig. 4.7).
    The primary aerosol types are

1. Sulphates. Most fuels used in cars, power stations and other industrial
   activity contain traces of sulphur and produce sulphur dioxide and hydrogen sulphide gas when burned. Natural processes like volcanic eruptions
   and the burning of biomass also produce sulphurous gases, although
   their contribution is estimated at only about 20% of anthropogenic
   sources. These gases react with other atmospheric species to produce
   solid (mostly ammonium sulphate) and liquid (sulphuric acid) particles
   in the atmosphere.

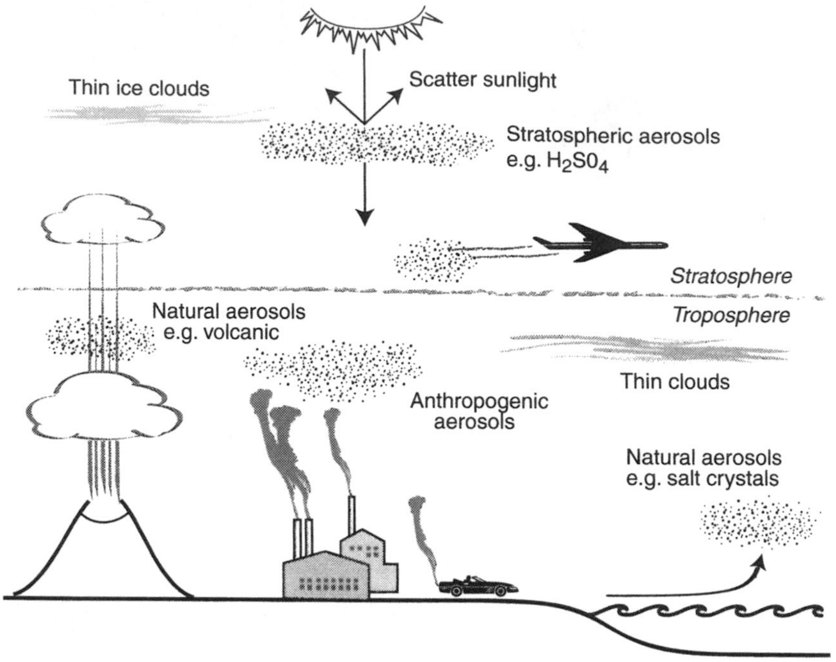

**Fig. 4.7**
Aerosols are ensembles of tiny (<1 μm) airborne particles of solid or liquid that, although often invisible to the eye, are ubiquitous in the atmosphere, and are found all the way up to the mesopause. They may be of human or natural origin, as illustrated. Stratospheric aerosols tend to be long-lived, and have an important effect on the albedo of the Earth, tending to reduce global warming. Tropospheric aerosols tend to attract water and grow, until they form cloud droplets, and eventually precipitate out as rain, hail, or snow.

2. Sea-salt particles, lifted by the wind as sea-spray evaporates.
3. Fine dust, raised by winds from deserts and other dry plains.
4. Carbon particles (soot) and other organics form the bulk of the visible part of smoke plumes from combustion processes of many kinds.

Sulphates and salt are soluble in water so these aerosols play a key role in the growth of cloud droplets (see Sections 4.6 and 4.7). Sulphate aerosols are particularly good at scattering visible light, and the buildup of these in the stratosphere is a major factor in global change, tending to work in the opposite direction to greenhouse gas accumulation. Dust and soot, on the other hand, absorb thermal infrared radiation and tend to heat the atmosphere, resulting in a net warming of the climate system.

The precise chemical compositions and optical properties of aerosols are highly variable; there are many different kinds of pollution and each may contribute in a different way. Some commonly found aerosols, such as mineral dust or weakly absorbing organic species, can either cool or warm the atmosphere depending on their particle size distribution and their chemical composition, which is so variable it affects the ratio of absorption to scattering. Location on the globe is also important, since prevailing weather conditions affect aerosol formation and behaviour. Stratospheric aerosols have long lifetimes, measured in years and even decades, but tropospheric aerosols tend to survive only for days or possibly weeks, resulting in large spatial and temporal variability over the surface of the Earth.

It is recognized that aerosols play an important part in the maintenance and evolution of the present climate, but also that their role is one of the most difficult to quantify and to model accurately, as further discussed in Chapter 10.

## Further reading

*A Short Course on Cloud Physics*, R.R. Rogers and M.K. Yau, Butterworth-Heinemann, 1996.
*The Physics of Clouds*, B.J. Mason, Oxford University Press, Oxford, 1971.
*Climatic Effects of Aerosols and Clouds*, K. Ya Kondratyev, Springer-Praxis Series in Atmospheric Physics & Climatology, 1999.

## Questions

1. What is meant by the *stability* of a parcel of air, with respect to its vertical movement? Define *potential temperature* and explain its importance in considering the stability of the atmosphere. Estimate the potential temperature $\theta$ at the tropopause assuming that the pressure there is 100 mb and the surface temperature $T$ is 290K, stating any assumptions that you make.
2. Obtain an expression for the entropy $S$ of a mass $m$ of unsaturated air, at pressure $p$ and temperature $T$. Derive the relationship between potential temperature $\theta$ and $S$. Describe how entropy varies with height under conditions of stability or instability.
3. Describe qualitatively how the temperature of a parcel of moist air will vary with height if it is forced to rise. Find an expression for the temperature lapse rate $-dT/dz$ in an atmosphere saturated with water vapour in terms of the lapse rate for a dry atmosphere, the molar latent heat of water, and the saturation vapour pressure.
4. Given the result of the previous question, describe how knowledge of the temperature profile can be used to infer the position of clouds. What factors affect the formation and growth of clouds? Comment on their relative importance in the formation of rain in the Earth's atmosphere in temperate latitudes.
5. A water droplet growing by condensation in a cloud of supersaturated air experiences a rise in temperature due to the release of latent heat. Find an expression for (a) the temperature of the drop as a function of its radius, $r$, and the thermal conductivity of air, $\lambda$, and (b) the rate of increase in the radius of the drop as a function of $D$, the diffusion coefficient of water vapour in air.
6. Describe briefly the contribution of each of the following to the formation and growth of droplets in a cloud and the eventual production of rain:

   (i) Homogeneous nucleation;
   (ii) airborne sea salt crystals;
   (iii) man-made pollution;
   (iv) airborne dust and smoke;
   (v) electrical charge;
   (vi) coagulation.

7. Explain the use of Kelvin's formula and Raoult's Law to obtain an expression for the critical radius for growth of a cloud droplet. How can this formula can be improved by the inclusion of van't Hoff's factor, and what physical effect is being allowed for by this modification?
8. Sketch the *Kohler curve* and obtain expressions for the critical radius above which a droplet will grow. Indicate this point on the diagram. For a droplet containing $10^{-17}$ kg of NaCl, the constants $a$ and $b$ have values of $6 \times 10^{-3}$ $\mu$m and $2 \times 10^{-5}$ $(\mu m)^3$, respectively. Calculate the approximate critical radius of the droplet, and the minimum level of supersaturation of the surrounding air required for droplet growth.

# 5 Ocean and climate

## 5.1  Introduction

Recapping the reasons why the ocean is an important component of the climate system:

1.  The ocean has a key role in the Earth's energy budget, primarily by contributing to the massive poleward transport of heat that is required to balance the latitudinal gradient in mean solar input. This meridional transport of energy is divided roughly equally between the atmosphere and the ocean, and if the ocean were not present or if its circulation slowed down the polar regions would be much colder, and the equatorial regions much warmer, than currently.
2.  The detailed balances in the hydrological cycle (the continuous cycling of water in all three of its phases between surface, ocean, and atmosphere) determine many of the most important thermodynamic properties of the atmosphere, including the vertical temperature profile, the heat capacity, and the factors that control cloud droplet growth and precipitation.
3.  The two most important of the 'greenhouse' gases that are responsible for the infrared opacity of the atmosphere—water vapour and carbon dioxide—are both in dynamic balance with large reservoirs in the ocean. The ocean is also a large thermal reservoir, and has the effect of delaying global warming due to its thermal inertia.

The ocean resembles the atmosphere in several basic ways, since it is also a fluid on a rotating sphere and is governed by a corresponding set of dynamical equations. However, there are huge differences in density, and hence heat capacity, and in the characteristic time and length scales of their dynamical behaviour. The density of air at sea level is about 1/1,000 of that of water, but it moves around the globe more quickly, by about the same factor, which is basically why atmosphere and ocean have a roughly equal role in global heat transport.

The radiative transfer of energy, which dominates in the atmosphere, is much less important in the ocean, because water is opaque at infrared wavelengths. The direct heating effect of the Sun is limited to the top few metres of water (the 'mixed layer'), where the radiation is absorbed and vigorous mixing occurs due to wind and waves. Below that, over a variable distance of typically several hundred metres, or about 10% of the average depth of 4 km, the temperature declines steadily. In the deep water, the temperature is nearly constant, declining only very slowly to reach typically a few degrees centigrade above freezing at the bottom.

Apart from temperature, the factor that most affects the oceanic structure and its global circulation is the concentration of salt. Salinity, like temperature, is highest near the top of the ocean because of the effect of evaporation (however, near the coast the influx of fresh water from rivers may reverse this tendency). Like temperature, salinity declines in the so-called thermocline region beneath the mixed layer (Section 5.4), and then becomes nearly constant with depth over most of the deep ocean, increasing slowly down to the bed.

The warm, high-salinity upper layer is not only mixed by the effect of the wind; the drag exerted by the moving air impinging on the water surface also produces the main top-level ocean currents. In the depths below the thermocline, the influence of the wind is not felt and the sluggish *thermohaline* circulation is controlled by density differences due to temperature and salinity contrasts. It is this that carries most of the energy around the globe and maintains the climate we know today. In order to understand this crucial process and monitor any changes, we need regular, extensive measurements in the deep ocean that unfortunately are difficult to obtain.

## 5.2 Ocean measurements

A major problem for the gathering of data on the global distribution of temperature and salinity in the ocean is the fact that only the *surface* of the ocean can be monitored from space, because the ocean is practically opaque to all but acoustic radiation. Thus, many of the advantages that satellite remote sensing has brought to atmospheric studies are not available to the oceanographer, and data on conditions below the surface is sporadic and difficult to acquire. Deep water temperature and salinity measurements mainly rely on direct sounding from ships (Fig. 5.1). These can trail cables with thermocouples attached to measure temperature and conductance meters to estimate salinity. For more accurate salinity measurements, clusters of sample bottles are lowered and filled at various depths. Upon retrieval, the salinity can be obtained from the specific gravity of the

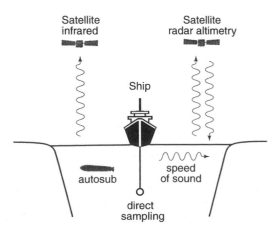

**Fig. 5.1**
Methods for monitoring the temperature and composition (especially salinity) of the ocean. Satellites can measure the temperature of the surface, and its topography; ships, including experimental mini-submarines, use direct sampling; advanced techniques are being developed that use shore-to-shore speed of sound measurements to infer temperature.

**Fig. 5.2**
The autonomous research vessel Autosub, with the covers removed from the sensor bay at the front and the propulsion unit at the rear. Most of the 7 m length is occupied by batteries. (Natural Environment Research Council, NERC).

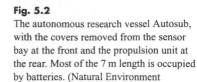

seawater at room temperature, using standard tables. Obviously global coverage is not achieved as, except for a few research vessels, data acquisition is limited to measurements of opportunity on the major sea routes.

Robotic vehicles, of which *Autosub* (Fig. 5.2), is a recent and successful example, provide a means to monitor the subsurface ocean using electronic sensors to measure temperature, salinity, and other variables. The route can be programmed into an onboard navigational computer, and the vessels can be launched from a ship or from land and left to function autonomously using their on-board battery power supply. At present they have a range of about 500 km, rather too short to serve as the fleet of 'underwater satellites' envisaged for the future monitoring of ocean conditions, but ideal for specific missions, especially to hostile locations. In one campaign, Autosub was deployed beneath the floating ice shelves in Antarctica, which form where polar glaciers meet the sea. These introduce vast amounts of fresh water in the form of melted snow and ice into the oceans, reducing the salinity and affecting the global circulation (see Chapter 10). Currently, about $1,000$ km$^3$ of fresh water are thought to melt into the ocean from the underside of Antarctic ice shelves each year, and measurements suggest that this process may be speeding up, although data on exactly what is happening and what processes are changing is very sparse. Autosub used temperature and salinity sensors to measure the rate of melting along the underside of the ice, the first time this had been done. It also used an acoustic Doppler profiler to measure ocean currents, and sonar to map the topography of the sea bed and the bottom surface of the ice sheet. A sample chamber on board collected water from beneath the shelf for later analysis of its detailed composition in the laboratory.

[51] The ATOC Consortium, Ocean climate change: Comparison of acoustic tomography, satellite altimetry, and modeling, *Science*, **281**, 1327–1332, 1998.

The ATOC (Acoustic Thermometry of Ocean Climate) project[51] (Fig. 5.3) makes use of the fact that sound travels faster in warm water than in cold water. The ocean acts like an acoustic wave guide and can carry sounds over very long distances, with the travel time of a pulse giving information on the average temperature between the source and the receivers placed around an ocean basin. Recent pilot studies to develop the technique used two sources, one near the California shoreline and one in Hawaii, with receivers at various locations in the North Pacific. At present it is possible to detect signals over a range of 1,000–2,000 km before they become lost in the natural background noise of the ocean, due mainly to scattering by eddies. The baseline for testing climate models needs to be greater, at least 3,000 km and preferably up to 10,000 km.

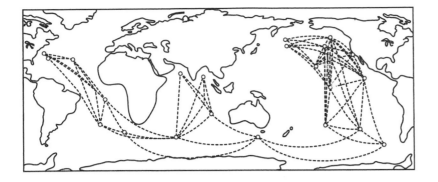

**Fig. 5.3**

A proposed scheme for global monitoring of ocean temperatures by acoustic thermometry, using a range of sources and receivers. In a pilot project, signals were transmitted from California to Hawaii and also to New Zealand, a distance of 10,000 km. (University of California.)

The data analysis approach uses Wilson's formula[52] for the velocity of sound $c$ (m s$^{-1}$) as a function of temperature $T$ (°C), pressure $p$ (bars) and salinity $S$ (‰):

$$c = 1449 + 4.6T - 0.055T^2 + 0.0003T^3 + (1.39 - 0.012T)(S - 35)$$
$$+ 0.0017p.$$

[52] W.D. Wilson, Equation for the speed of sound in seawater. *J. Acoust. Soc. Amer.*, **32** (N 10), 1357, 1960.

For the ranges of $S$ and $T$ found in the ocean, salinity has relatively little effect on sound velocity, compared to temperature, so climatological information on the value of $S$ and $p$ are sufficient to allow the mean temperature of the path, $T$, to be treated as the only unknown. For climate monitoring, detection of changes as small as around 0.005K yr$^{-1}$ at 1 km depth are needed, so $c$ has to be measured to significantly better than about 1 part in $10^5$.

Fortunately, one of the most important aspects of ocean behaviour—the exchange of heat and moisture with the atmosphere—can be monitored globally from space. A key parameter is the sea surface temperature (SST), which is accessible to infrared radiometers on satellites if they make measurements at 'window' wavelengths where the atmosphere is mostly transparent (see Chapter 9). The ocean surface has temperature contrasts of around 1K on a regional scale, due to the presence of currents and variable mixing and evaporation resulting from interaction with the wind field. Global measurements of SST reveal not only patterns of evaporation and cooling, but also circulation features such as El Nino (see Section 10.5.3) and the Gulf Stream (Section 5.6) to be monitored as they change from day to day and seasonally.

Radar transmitters and receivers on satellites can obtain the distance between the satellite and the sea surface from the round-trip travel time of microwave pulses bounced from the spacecraft to the sea surface and back to the spacecraft. If the altitude of the satellite is determined separately by accurate tracking using transponders, maps of the topography (i.e. the fluctuations in height) of the ocean surface are obtained. For this to be useful in oceanography it is necessary to have simultaneous knowledge of the *geoid*, which is the mean figure of the Earth as manifest by its gravitational field, and is where sea level would lie if the ocean were homogeneous and stationary.[53] The accuracy with which the geoid is known is steadily improving due to gravity field measurements also made by satellites.

[53] The geoid itself is a complex shape, because the solid Earth is neither homogeneous nor smooth; the undulations produced by ocean dynamics and density variations are around 100 times smaller than the deviations in the geoid.

The pressure forces associated with major currents, tidal forces, and surface gravity waves, manifest themselves in the measured surface height variations. Recent satellite systems like TOPEX achieve an accuracy of about 2 cm, which can be compared to global variations of around ±20 cm. These are due mainly to dynamical effects, in which the slope in the ocean surface produces a horizontal force to balance the Coriolis force corresponding to the surface current.[54] A slope of 1 m in 100 km is typical of strong currents. A careful analysis of the topography data allows changes in the mean current, and the strength of large eddies, to be followed. Also, since the measurements are absolute, a long series of observations can be used to look for trends in global sea level; mean rises of as much as 90 cm are predicted to occur in the next half-century as the ocean expands in response to global warming (see Fig. 1.17).

In addition to 'time of flight' measurements for altimetry, the details of the shape of the returned radar pulses can be analysed to obtain valuable information on wave height and surface roughness. The latter can be interpreted in terms of the wind near the ocean surface, and hence of the currents in the upper ocean, since these are driven by the wind, as discussed below (Section 5.6).

[54] This is a form of *geostrophic* balance; we discuss what this means, and recap the nature of the Coriolis force, in Section 5.6 below.

## 5.3  Composition of the ocean: Salinity

The ocean is, of course, composed primarily of $H_2O$, but, as in the atmosphere, minor constituents have a key role to play in controlling the climate. In Section 1.4.3 we discussed how the ocean can take up, sequester, or release, large amounts of carbon dioxide, thus regulating the atmospheric greenhouse, and how oceanic life (especially the minute single-celled plants known collectively as phytoplankton) can also play an important part in the climate system, just as land-based life does. However, the most crucial aspect of ocean composition is its salinity, the dissolved salts that have a large effect on water density and hence in the ocean circulation and its role in global heat transport.

The ocean is salty because of the weathering of rocks in the early stages of the formation of the oceans and continents. The ocean is probably slowly becoming saltier, on geologic time scales, as rivers bring additional dissolved material to the sea. Although small amounts of salt are removed from the ocean, for example by the evaporation of spray to form tiny crystals light enough to be carried by the wind (which may then become important in cloud nucleation processes, see Chapter 5), this is probably outweighed by the influx, and salt slowly builds up.

Sea water is about 3.5% salt by weight on average, which means the total quantity of salt if extracted could form a layer 45 m thick over the entire Earth, or 153 m thick over the land area. The salt in seawater is composed primarily of sodium chloride (~85%), with smaller amounts of magnesium chloride, magnesium sulphate and calcium carbonate, along with many trace constituents (Table 5.1). One of the remarkable aspects of ocean salinity is how constant the relative ratios of these components are in sea water, no matter where in the oceans one looks nor what the value for total salinity. This indicates that the oceans are well mixed on the time scale of salt input.

**Table 5.1** The main salts present as minor constituents in the ocean, expressed as percentages of each radical by mass. In addition to these, there are a number of trace elements, for example, manganese (Mn), lead (Pb), gold (Au), iron (Fe), iodine (I) in concentrations of parts per million or less. Seawater also contains small amounts of dissolved gases (nitrogen, oxygen, hydrogen, etc. as well as carbon dioxide).

| | |
|---|---|
| Chloride (Cl) | 55.04% |
| Sodium (Na) | 30.61% |
| Sulphate (SO₄) | 7.68% |
| Magnesium (Mg) | 3.69% |
| Calcium (Ca) | 1.16% |
| Potassium (K) | 1.10% |

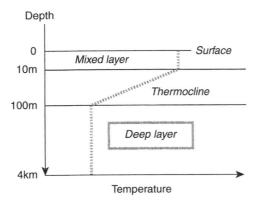

**Fig. 5.4**
Schematic of the mean vertical temperature structure of the ocean (note the nonlinear depth scale). The depths of the mixed layer and of the thermocline shown are typical, but vary substantially with latitude, season, and surface wind field (cf. Fig. 5.6.)

Total salinity varies from about 3.2% to 3.8% in open-ocean surface waters, depending mainly on the balance between precipitation and evaporation in different regions. The general pattern is one of low salinity near the equator (high evaporation, but very high precipitation) and near the poles (low evaporation and high precipitation) with maxima at mid-latitudes. The run-off of relatively fresh water from the continents as rivers is also a key factor that varies seasonally. In the ocean near large sources of fresh water, such as the mouths of major rivers or melting ice masses, the salinity can fall to values well below 3%. On the other hand, the salinity in the deepest parts of the global ocean is much less variable, staying close to 3.5%. As a result, the mean density of the global ocean as a whole is about $1.035 \text{kg m}^{-3}$.

## 5.4 Vertical and latitudinal structure of the ocean

Figure 5.4 shows the three-layered approximation to the mean vertical structure of the ocean described in the introduction to this chapter. The uppermost few metres are the *mixed layer,* in which the ocean is stirred by the wind and waves and the temperature is essentially constant. This is separated from the *deep* (or *abyssal*) layer, which constitutes around 98% of the ocean and which is decoupled from the direct influence of Sun and wind by the *thermocline* region, in which temperature declines steadily, becoming nearly constant again in the deep layer.

Salinity follows a similar pattern to temperature, with relatively high values in the mixed layer, a steady decline going downwards through the thermocline, and remaining nearly constant through the deep layer. In actual measured profiles, like the examples in Fig. 5.5, the temperature declines slowly, and the salinity increases slowly, down towards the seabed. Both these trends are a consequence of the fact that the density of water increases with decreasing temperature (below 4°C) and with increasing salinity (see Section 5.5).

Over the years, a large number of individual profiles of temperature and salinity has been collected and used to construct a global picture of the ocean. In addition to seasonal and random fluctuations, this reveals a distinct latitudinal pattern. Figure 5.6 is a simplified picture of this, showing how the depths of the mixed layer and the thermocline are observed to vary from north to south in the Atlantic. The key feature is that both of the near-surface layers essentially vanish near the poles, where the deep layer extends to the surface at latitudes near the edge of the polar icecaps. This is in fact where the deep water forms, before flowing downwards and equatorwards.

The strong downwelling that occurs uniquely in the polar oceans is the result of the generally low temperatures, further depressed by strong winds to the point where sea ice begins to form. The ice that forms is nearly fresh, leaving an increased salt content in the remaining water and reducing its freezing point. This cold, salty brine is extremely dense, and sinks to the sea bottom, where it fills the basins at the bottom of the polar seas, and eventually the rest of the deep ocean. This water eventually returns to the surface at lower latitudes, often centuries later, since over most of the rest of the global ocean, slow upwelling is the norm (Fig. 5.6).

The temperature gradient in the thermocline and its depth at any particular location depend on a number of factors, including the surface winds and the solar heating, but overall are dominated by the balance between vertical motions (advection) and diffusion due to small-scale eddies. The rate of change of temperature in the mixed layer can be estimated from a simple equation expressing the balance between diffusion and advection in the

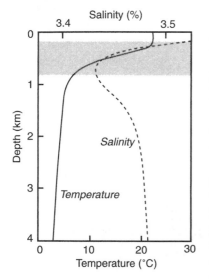

**Fig. 5.5**
Measured temperature and salinity profiles, typical of the mid-latitude ocean.

**Fig. 5.6**
A cross-section of the Atlantic Ocean, showing the extent of the mixed layer and the thermocline, and the formation near the polar caps of the deep water layer that makes up more than 90% of the global ocean by a combination of radiative and evaporative cooling, and increased salinity due to the separation of water as sea ice.

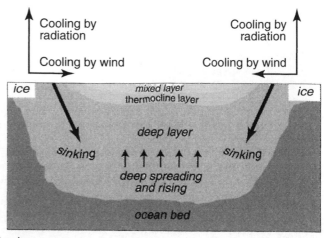

thermocline region, thus:

$$\frac{\partial T}{\partial t} = -w\frac{\partial T}{\partial z} + \kappa\frac{\partial^2 T}{\partial z^2},$$

where $\kappa$ is the diffusion coefficient and $w$ the vertical velocity or advection rate that applies to the upwelling that occurs over most of the global ocean.[55] In the steady state the left-hand side of the equation is zero, and the resulting differential equation can be solved. It is straightforward to show by substitution that a solution exists in the form

$$T = T_0 + T_1 e^{-z/z^*},$$

where $T_0$ and $T_1$ are constant and $z^* = \kappa/w$ is a scale depth for the temperature profile in the ocean. This model fits the average behaviour of the real ocean if $z^*$ has a value of around 500 m.

For typical values of $\kappa$, this predicts mean global upwelling rates through the thermocline that are very small, $\sim 10^{-4}$ m s$^{-1}$. Mass is conserved by the sinking of cold, salty water in certain specific regions of the N. Atlantic and the Antarctic, as discussed above; because these sinking regions are much more confined in area than the general upwelling that occurs virtually everywhere else, their downwards vertical motions are much more rapid.

## 5.5 The oceanic equation of state

The equation of state describes the density $\rho$ of the ocean as a function of temperature, salinity and pressure:

$$\rho = \rho(T, S, p).$$

For fresh water, $\rho$ decreases with increasing temperature above 4°C, and increases below. It increases monotonically with increasing salinity; both dependences are non-linear (especially that with temperature) and also vary slightly with pressure. Unlike fresh water, which has its maximum density at 4°C, typical seawater containing 3.5% dissolved salt continues to get denser as it gets colder, and the coldest water is found at the bottom of the ocean.

Pressure increases by 1 bar approximately every 10 m of depth, and so reaches values in excess of 400 bars at the ocean bed. For pressures as large as these, the compressibility of water is significant, changing the density from 1,028.1 kg m$^{-3}$ at the surface to 1,032.8 at 1 km and 1,046.4 at 4 km (all for fixed values of salinity and temperature, $S = 35$ psu and $T = 0$°C). For comparison, the surface value of 1,028.1 would fall to 999.8 kg m$^{-3}$ if all salinity were removed while keeping the temperature constant at $T = 0$°C. These differences are crucial in determining the global circulation of the deep ocean.

Unlike the atmosphere, there is no simple law derived from first principles that describes the dependence of density on temperature, composition, and pressure. We must depend on measurements to establish empirical plots and formulae that describe these relationships.[56]

[55] For a fuller discussion of the vertical propagation of heat in the ocean, including the formal derivation of this equation and its solution, see M.I. Hoffert et al., *J. Geophys. Res.*, **85**, 6667–6679, 1980.

[56] As an illustration of this, the following is an example of the equation of state from an advanced climate model. Here density $\rho$ is expressed as a function of potential temperature $\theta$ (°C) (the temperature the water would have if raised adiabatically to the surface, cf. Section 4.2), salinity $S$ (parts per thousand, ‰) and pressure $p$ (bars): $\rho = p_1/(p_2 + 0.7028 p_1)$, where $p_1 = p + 5884.81 + \theta(39.80 + \theta(-0.31 + 0.00042\theta)) + 2.61S$, and $p_2 = 1747.4508 + \theta(11.5158 - 0.04633\theta) - S(3.8542 + 0.01353\theta)$.

This complex expression has terms in $\theta^3$ as well as cross-terms involving the product of all three variables, reflecting the complexity that is found to be necessary to make sophisticated models of the ocean.

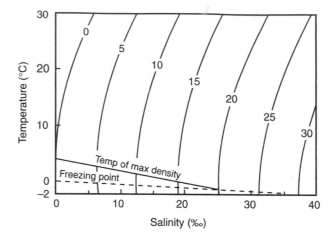

**Fig. 5.7**

A plot of the equation of state for seawater, showing density (expressed as the difference in parts per 1,000 (‰) from the value at $S = 0$ and $T = 0°C$) as a function of temperature and salinity at a fixed pressure of 1 bar, that is, near the surface. Note that (1) density is almost independent of temperature near the freezing point, (2) both the freezing point, and the temperature at which density is a maximum, decline with increasing salinity. Most of the water in the ocean would be found in the lower right-hand quadrant of this plot.

Note that the curves of constant density in Fig. 5.7 are nearly vertical, that is, parallel to the temperature axis, for low temperatures where the water is near freezing. Under those conditions, changes in $S$ have a more important effect on density than $T$ does. This reverses at higher temperatures, like those found in tropical waters near the surface. If salinity were the same everywhere, we would expect to find dense water subsiding near the poles and being replaced with warmer, less dense water from lower latitudes, which broadly speaking is the basis for the global circulation of the bulk of the ocean, as we shall see later in this chapter. It can also be inferred from the plot in Fig. 5.7, however, that if the polar oceans were significantly less salty than the tropics, for example, because the icecaps were melting and releasing fresh water, the salinity effect might reverse the density gradient and slow down, or even reverse, the circulation pattern. This is one of the possible scenarios for global climate change that we will examine in Chapter 11.

## 5.6 The general circulation of the ocean

The ocean circulation has two distinct components: the top few hundred metres, containing the mixed and thermocline layers, which is driven by drag between the water surface and the prevailing winds, and the deep ocean, which is driven by differences in density resulting from temperature and salinity gradients. In both cases, the long-term average motions (i.e. with waves and turbulence averaged out) are determined mainly by the balance between the main driving force (wind drag and density gradients, respectively) and the *Coriolis force* that enters as a result of the Earth's rotation.

## 5.6.1 The Coriolis[57] Force

The Coriolis force is often introduced in terms of its effect on objects moving across a rotating turntable or roundabout. The size of the effect is quite small in such cases, however, and most of us have little direct experience of the Coriolis force at work. It is important to realize, however, that it is *extremely* important when applied to global-scale motions in the atmosphere and ocean, because of the much larger spatial scales involved compared to the everyday experience of merry-go-rounds etc. As just noted, the mean motions of the upper layers of the ocean can be understood to first order as a balance between the motion due to wind drag and the resulting Coriolis force. At depths of more than a few hundred metres, where the effect of the wind is no longer felt directly, the ocean moves in response to pressure gradients produced by buoyancy forces, themselves produced by temperature and salinity gradients, and these motions are also subject to important modifications by the Coriolis force. Other key factors are drag at the bottom of the ocean, and, of course, the presence of the solid boundaries at the coastlines.

To see how the Coriolis force arises, consider a body of mass $m$ at the equator that is stationary with respect to the solid Earth (Fig. 5.8). Since the planet rotates with angular velocity $\Omega$ radians per second, $m$ has a linear velocity $R\Omega$ in the $x$-direction (eastwards) in an external reference frame, fixed with respect to the stars. If $m$ now moves polewards on the surface of the globe, along a meridian and without friction, it still has velocity $v$ but the underlying Earth has a smaller velocity $r\Omega$, corresponding to the smaller distance $r$ from the axis of rotation (Fig. 5.9). The mass $m$ thus appears, to an observer on the Earth, to be moving eastwards with a velocity $u = (R - r)\Omega$ relative to the surface. If the mass continues to move polewards, with a constant velocity in the $y$-direction of $v$, it will be observed to accelerate eastwards; the force responsible for this acceleration is the Coriolis force $F_C$.

The magnitude of $F_C$ can be obtained from considerations of the angular momentum $I = m\Omega r^2$ of the mass $m$. For conservation of $I$, $F_C$ must exert a torque equal to the rate of change in the angular momentum, so we have

$$\frac{\mathrm{d}I}{\mathrm{d}t} = \frac{\mathrm{d}}{\mathrm{d}t}\left(m\Omega r^2\right) = 2m\Omega r\frac{\mathrm{d}r}{\mathrm{d}t} = rF_C.$$

For a poleward velocity $v$ we also have

$$\frac{\mathrm{d}r}{\mathrm{d}t} = \sin\phi\frac{\mathrm{d}y}{\mathrm{d}t} = v\sin\phi,$$

where $\phi$ is latitude, so

$$F_C = 2m\Omega v\sin\phi.$$

This derivation gives the same equation regardless of whether the object is in the northern or southern hemisphere; the force on a poleward-moving mass is perpendicular to the motion and towards the East in either hemisphere. For motion away from the pole in either hemisphere, the object is accelerated westwards. $F_C$ is greatest at the pole, and zero at the equator.

[57] Gaspard Gustave de Coriolis (1792–1843) taught mechanics at the Ecole Polytechnique in Paris from 1816 to 1838, and wrote, among other works, 'Sur les équations du mouvement relatif des systémes de corps'. (*J. de l'Ecole Royale Polytechnique*, **15**, 144–154, 1835) in which he showed how the equations of motion could be modified for use in a rotating coordinate frame by including an additional acceleration.

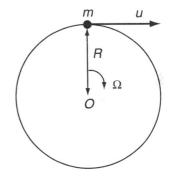

**Fig. 5.8**
The circle represents the equator of the Earth, of radius $R$ rotating with angular velocity $\Omega$, which we are viewing along the rotation axis from the south pole. A body of mass $m$, stationary with respect to the surface, has a linear velocity $u = R\Omega$ in the $x$-direction (eastwards), and an angular momentum of $I = mvR = m\Omega R^2$, by virtue of the Earth's rotation.

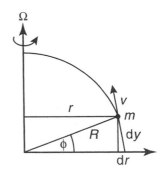

**Fig. 5.9**
The mass $m$ moves polewards from the equator to latitude $\phi$, a distance $\mathrm{d}y$, in time $\mathrm{d}t$, with constant velocity $v$. It experiences a Coriolis force perpendicular to $v$.

A similar analysis gives the expression for the Coriolis force on a body that is moving across the surface of the Earth in the zonal direction, parallel to the equator. Consider steady motion with velocity $u$ towards the East, at latitude $\phi$ which corresponds to a distance $r$ from the rotation axis. We have

$$F_C = m \left( \Omega + \frac{u}{r} \right)^2 r = m\Omega^2 r + \frac{mu^2}{r} + 2m\Omega u$$

for the total force on $m$, where $u/r$ is the extra angular velocity of the body due to its motion relative to the surface of the Earth. Of the three terms on the right, the first is the centrifugal force on the body at rest, the second is the addition to that by virtue of its motion, which is small compared to the previous term and can be neglected, and the third term is the Coriolis force. The Coriolis force has components acting vertically and in the N–S direction towards the equator. The latter is what interests us and is given by:

$$F_{N–S} = 2m\Omega \sin \phi u.$$

This derivation also gives the same equation regardless of whether the object is the northern or southern hemisphere; the force is perpendicular to the motion and to the right (equatorwards) in the northern hemisphere and to the left (also equatorwards) in the southern hemisphere, and vanishes at the equator.[58]

### 5.6.2   Deep ocean dynamics: the thermohaline circulation

The deep ocean layer constitutes more than 90% of the mass of the ocean, that is, all but the top few hundred metres of depth on average. We saw in Section 5.4 that the deep layer can be considered to extend all the way to the surface near the poles, where the surface waters are particularly cold and dense, resulting in strong downwelling towards the ocean floor. This rapid sinking motion is unique to the polar regions, and indeed appears to be focussed in relatively small areas in the extreme N. Atlantic near Greenland in the north, and in the Weddell Sea in the south. Here, particularly cold and dense surface waters are found due to:

(1)  the large excess in radiative cooling over solar heating at high latitudes;
(2)  strong evaporative cooling by the strong and persistent winds found there;
(3)  freezing of some of the water, causing brine enrichment in the remainder, since sea ice is relatively fresh, at 5–7‰ of salt compared to an average of 35‰ for sea water.[59]

Over the rest of the globe, the vertical motions in the ocean are generally much weaker, and tend to be upwards, balancing the strong downward flow in the Arctic and Antarctic. Thus, the combination of the thermal structure of the ocean and its salinity field give rise to a slow, global, overturning motion known as the *thermohaline circulation,* so called because it is driven by density variations primarily induced by temperature (thermo) and salinity (haline) differences between water masses. Figure 5.11 shows a schematic

[58] Picture, for instance, the situation where air is flowing in all four directions simultaneously, as happens in a region of low atmospheric pressure (Fig. 5.10). The deflection of this flow by the Coriolis acceleration causes the low pressure system to spin up in the anticlockwise direction (in the N. hemisphere, clockwise in the S.), producing the features known as *cyclones* that are familiar on everyday weather charts. The high-pressure equivalent rotates in the opposite sense and is called an *anticyclone*.

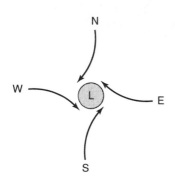

**Fig. 5.10**
When a low-pressure system develops in the atmosphere, air flows into it from all points of the compass, and each is deflected by the Coriolis force as shown (for the N. Hemisphere), resulting in rotation of the air mass and formation of a cyclone.

[59] This process occurs in reverse when polar icecaps melt, and become a source of relatively fresh, low-density water, like melting snowfields on land or heavy rainfall. One of the concerns in a world undergoing global warming is that the fresh water released by melting ice caps will reduce the salinity, the density, and hence the downwelling in the polar seas, and so slow down or even reverse the deep ocean circulation (see Section 10.5).

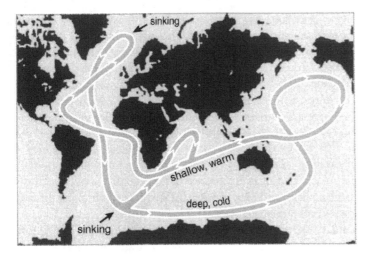

**Fig. 5.11**
The mean behaviour of the thermohaline circulation of the deep ocean forms a global 'conveyor belt' that is the primary route for transportation of heat in the ocean. Downwelling occurs almost uniquely in the N. Atlantic and the Weddell Sea in Antarctica, matched by global upwelling that occurs almost everywhere else, but is greatest in the Indian and Pacific oceans.

representation of this, often referred to as the 'global conveyor belt' for its role in transferring heat, salt, and tracers around the globe.

This figure is greatly simplified (in particular, mixing is not represented), but does show several key features of the deep ocean circulation. As noted already, communication between surface and depths (or 'bottom water formation', as oceanographers say) in the northern hemisphere occurs exclusively in the isolated polar regions of the Atlantic. The cold, dense water flows southwards as it sinks, forming the lower limb of the northernmost branch of the circulation. There is no similar behaviour in the north Pacific or the Indian Oceans, since the former is comparatively fresh and the latter comparatively warm. Both are therefore considerably less dense than the north Atlantic, and are characterized by general upwelling.

Bottom water formation in the southern hemisphere occurs in the southernmost reaches of the Southern Ocean, and in particular in the Weddell Sea, which experiences salinity enrichment both as a result of evaporation and as a result of the formation of sea ice. These waters are the densest anywhere on the globe, and sink to great depths, where they feed into the Pacific and Indian Oceans. After rising (accompanied as always by some mixing), these streams reunites to form the shallow branch of the Atlantic current, which heads northwards via the west coast of Africa and the Gulf of Mexico.

The motion in the deep ocean is extremely slow, and predominantly adiabatic. This means it is heat conserving (as we assumed when deriving thermodynamic expressions for the atmosphere, although in fact the approximation is better when considering the ocean), and salt-conserving. This makes it possible to work out when water last encountered the surface from its temperature and salinity, and its content of soluble *tracers*. These are minor constituents with origins that are known in space and time, and include radioactive species like tritium ($^{3}$H) and radiocarbon ($^{14}$C), produced from nuclear bomb tests, and chlorofluorocarbons, the man-made chemicals that play a large part in ozone depletion, which have been increasing with time in a known way. Tracer studies show that the 'conveyor belt' concept is valid enough to be useful, despite some mixing and some general upwelling everywhere outside the polar seas. For instance, water starting out from near the surface (just below the mixed layer) in the mid-Pacific can be followed to the Indian Ocean, and from there into the north Atlantic, and then back into the Pacific, passing 2 or 3 km below its starting point several decades after it left. The tracer content of some of the water in the very deep (abyssal) regions of the ocean shows, in some cases, that it has not been in contact with the surface and the atmosphere for *centuries*.

The task of computing the quantitative dynamical behaviour of the ocean is made difficult by the complicated and implacable boundaries formed by the shorelines. However, we can obtain some simple properties of the thermohaline circulation in the interior of the ocean by looking at the balance between the density gradients and the Coriolis force, which we already noted are the two dominant influences.

The cold, dense water sinking in the N. Atlantic reaches the bottom of the ocean and moves south, driven by the pressure gradient and channeled by the land masses on either side. However, as it moves, the Coriolis acceleration forces the flow towards the direction at right angles to its initial movement. In the northern hemisphere, this force is to the right, so for the water that sank in the N. Atlantic, it moves towards the Americas as it flows south, until an east–west pressure gradient builds up to balance it (Fig. 5.12).

In the interior of the ocean, frictional forces can be neglected and the steady state circulation represents the balance between the pressure gradient force and the Coriolis force. This is *geostrophic* flow, in which the water moves along contours of constant pressure (isobars), with high pressure on the left in the southern hemisphere, on the right in the northern. The pressure differences are reflected in variations in the height of the sea surface of up to a metre or more, which can be observed using laser altimeters on satellites orbiting above.

From Fig. 5.12, we see that the east–west pressure gradient is

$$\frac{\Delta p}{\Delta x} = \rho g \frac{\Delta z}{\Delta x} = \rho g \tan \theta,$$

so equating this to the Coriolis force gives

$$\rho g \tan \theta = \rho 2 \Omega \sin \phi v$$

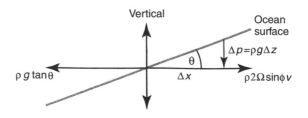

**Fig. 5.12**

A simplified concept of geostrophic flow in the deep water of the N. Atlantic. The initial pressure gradient, due to cold, salty water sinking in the extreme north, is towards the south (into the page), with America on the right. The Coriolis force, towards the west, is balance by an induced pressure gradient in which the ocean is higher on one side than the other.

whence

$$v = \frac{g \tan \theta}{2\Omega \sin \phi}.$$

The only unknown in this expression for the velocity is the slope of the pressure surface, $\theta$, relative to the local horizontal (i.e. the surface with a constant value for the acceleration due to gravity, $g$), and it can be applied at every drag-free level in the ocean, that is, away from the surface, the bed, or the coast. We can also write

$$\frac{\partial p}{\partial x} = \rho g \tan \theta = \rho 2\Omega \sin \phi v$$

so

$$v = \frac{1}{\rho 2\Omega \sin \phi} \frac{\partial p}{\partial x}$$

therefore

$$\frac{\partial v}{\partial z} = \frac{1}{\rho 2\Omega \sin \phi} \frac{\partial}{\partial z} \left( \frac{\partial p}{\partial x} \right) = \frac{1}{\rho 2\Omega \sin \phi} \frac{\partial}{\partial x} \left( \frac{\partial p}{\partial z} \right)$$

and, since $\partial p / \partial z = -\rho g$ by the hydrostatic relation,

$$\frac{\partial v}{\partial z} = \frac{-g}{2\rho \Omega \sin \phi} \frac{\partial \rho}{\partial x}.$$

This is called the *thermal wind equation*, because it was derived originally for the atmosphere where the horizontal density gradient depends only on the temperature. It relates the variation in height of the velocity in the $x$ (north–south) direction to the density gradient in the perpendicular (east–west) horizontal direction.

### 5.6.3 Near surface dynamics: Ekman transport and Sverdrup balance

Unlike the density-driven flow in the deep ocean, the motions near the surface are mainly produced by the effect of wind drag, transferring momentum from the atmosphere to the surface and near-surface layers of the water. The wind also generates surface waves that further affect the coupling between wind and water, and produce vertical as well as horizontal motions. These cause turbulent mixing of the top $\sim$10–100 m, producing the nearly iso-thermal mixed layer (described above in Section 5.4). Here, the involvement of frictional forces cause a departure from geostrophic flow; water moves across isobars from areas of high pressure to areas of low pressure. The layer in which the flow is non-geostrophic is known as the *Ekman* layer (more strictly, the *surface* Ekman layer, since there is also a 'bottom' Ekman layer corresponding to the vertical range of influence of drag by the ocean bed on the moving water).

In the simplest case where the horizontal pressure gradients due to sea surface slope and internal density differences can be neglected entirely, the

[60] Vagn Walfrid Ekman (1874–1954) and Harald Ulrik Sverdrup (1888–1957) were physical oceanographers of Swedish and Norwegian origin, respectively, who studied the dynamics of ocean currents.

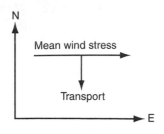

**Fig. 5.13**
Ekman transport in the northern hemisphere.

only force available to balance the wind stress is the Coriolis force associated with the large-scale movement of the water in the layer. Since this acts to the right of the motion in the northern hemisphere, the mean flow must be at right angles to the wind stress, to the left in the southern and to the right in the northern hemisphere (Fig. 5.13). This is known as *Ekman Transport*.[60]

Using the expression for the Coriolis force derived above, a simple estimate of the amounts of water moved by Ekman transport can be made as follows. First, we note that the surface stress $F$ exerted on the water is related to the wind speed $w$ by the empirical formula

$$F = C_d \rho_\alpha w^2,$$

where $C_d$ is a dimensionless constant called the drag coefficient, $\rho_\alpha$ is the air density and $w$ is wind speed. Setting the frictional force $F$ due to the wind equal to the Coriolis force gives

$$F = 2\Omega \sin \phi u \rho D,$$

where $\rho$ is the water density and $D$ the depth of the layer affected by wind drag (the Ekman layer), $\Omega$ is the angular velocity of rotation of the Earth, and $\phi$ is latitude. Defining the transport $Q$ as the product $uD$ we get

$$Q = \frac{F}{2\rho\Omega \sin \phi}.$$

For typical values of wind stress, this formula predicts that several tonnes of water are moved each second at right angles to the wind for each metre measured along the wind direction.

Along the west coast of South America, where the prevailing wind blows southward in summer, the ocean current will be away from the coast. This forces the upwelling of cold, deep water, which explains why the waters off the coast of Peru are colder than the average at that latitude. Similar behaviour is observed off the western coasts of California and Africa, but the South American case is the best known, because the prevailing winds there are subject to the El Niño–Southern Oscillation (ENSO) oscillation (Section 10.5.3) that causes warm water to shift across the Pacific in a pattern that has a marked effect on the climate on a global scale.

On a larger scale, the effect of the global mean wind field is to drive the circulation of the mixed layer at the top of the ocean in asymmetric circulation patterns known as *gyres*. These are illustrated in Figs 5.14 and 5.15. A simple version of the theory due originally to Sverdrup can be used to understand the formation of gyres, starting with the concept of the *vorticity* of a water parcel. Vorticity is a measure of the rotation of an object, and is defined in the general case as the curl of the velocity vector. For a solid body, this is twice its angular velocity $\omega$, with *anticlockwise* rotation representing *positive* vorticity. Ocean water has both *relative* vorticity $\zeta_r$ (relative to the solid Earth), and additional *planetary* vorticity $\zeta_p = 2\Omega \sin \phi$ due to the rotation of the planet, making the total or *absolute* vorticity $\zeta$ the sum of the two.

Vorticity is a useful concept because it is conserved in frictionless flow. In the ocean, as we have seen, fluid parcels move large distances meridionally

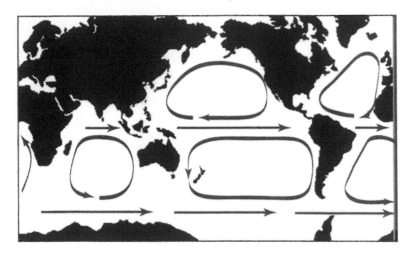

**Fig. 5.14**
A schematic of the circulation of the upper (wind-driven) part of the global ocean, showing its organisation into large gyres, either side of the equator.

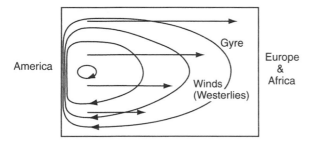

**Fig. 5.15**
A simple model of the mid-latitude north Atlantic ocean, showing the prevailing winds (Westerlies) parallel to the equator and increasing steadily to the north, and the ocean circulation (sub-tropical gyre) resulting from balance between wind drag, the Coriolis force, and friction with the coastlines. The main part of the gyre exhibits features of Sverdrup balance, while the return flow along the American coast is called the western boundary current.

(north–south) under virtually frictionless conditions, because the viscous forces are proportional to the velocity gradients, which are small. Since the planetary vorticity depends on latitude, the relative vorticity must also change in order to conserve the sum of the two. Since

$$\zeta = \zeta_r + \zeta_p = \text{constant}$$

we can write

$$\frac{d\zeta_r}{dt} = -\frac{d\zeta_p}{dt} = -\frac{d(2\Omega \sin\phi)}{dt}$$

$$= -2\Omega \cos\phi \frac{d\phi}{dt} = -2\Omega \cos\phi \, Rv$$

$$= -\beta v,$$

where $R$ is the radius of the Earth, $d/dt$ is the rate of change following a water parcel, and $\beta$ is a constant. This states that the rate of change of relative vorticity $\zeta_r$ is directly proportional to the north–south velocity $v$. Thus, in the northern hemisphere, a *northward*-moving water parcel acquires *clockwise* (negative) vorticity or spin because of the Coriolis effect. Conversely, it will acquire anticlockwise (positive) vorticity by moving south.

Now consider the effect on the vorticity of a water parcel of frictional stresses due to the wind. Figure 5.15 shows a simplified representation of the prevailing wind in the mid-Atlantic (the Westerlies), in which the eastward wind stress increases as we go north. For steady motion in a steady wind field, the force $F$ due to the gradient in the wind stress is

$$F = \rho h \frac{du}{dt} \approx \frac{u}{t}.$$

Consider how $u$ varies in the north–south direction $y$:

$$\frac{\partial u}{\partial y} = \frac{t}{\rho h} \frac{\partial F}{\partial y} = -\zeta_r$$

and how the relative vorticity $\zeta_r$ varies with time, following a parcel:

$$\frac{d\zeta_r}{dt} = \frac{d}{dt}\left(-\frac{\partial u}{\partial y}\right) = -\frac{1}{\rho h}\frac{\partial F}{\partial y},$$

where we have assumed that there is no east–west gradient in the wind velocity, again a realistic assumption in general, so that $\partial v/\partial x = 0$. We see that an *eastward* wind stress that increases going *polewards*, continuously adds a *clockwise* spin to the water column at every point.

If the water parcel moves south at a rate in which its increase of anticlockwise vorticity due to its decrease of latitude, and the gradient with latitude of the Coriolis force, just matches the clockwise input from the wind, we have the condition known as *Sverdrup balance,* expressed by

$$v = -\frac{1}{\rho h \beta}\frac{\partial F}{\partial y}.$$

For this vorticity balance to hold everywhere, *all of the water* in the subtropical gyre must have a southward component. This is clearly impossible, but in the real ocean we do find that in most of the gyre the flow is equatorwards, and it is only in a relatively narrow western strip that northward flow occurs (Fig. 5.15). Thus, Sverdrup balance applies approximately across most of the ocean and contributes to an understanding of the observed asymmetry of the circulation pattern.

The general pattern of surface currents in the ocean is well known from centuries of navigation and measurement. In the northern hemisphere, each of the ocean basins actually contains two counter-rotating gyres in equilibrium with the wind forcing. The southern hemisphere circulation is not a mirror image of that in the north, having a single, large gyre, because of the existence at higher latitudes of the strong Antarctic circumpolar current. The Antarctic is the only ocean where the water can travel right around the globe without meeting a continental barrier. Each of the gyres has the predicted asymmetry, with currents that are much stronger on the western boundary than on the eastern. These *western boundary currents* all have familiar names: the one in the north Atlantic is the Gulf Stream, which transports 100 million cubic metres of warm water per second in a stream about 100 km wide, off the east coast of North America.

## 5.7  Ocean circulation and climate change

The description in this chapter of the general structure and circulation of the ocean deals mainly with the steady state, and its role in sustaining the present-day climate. There remain questions of the *stability* of the coupled atmosphere–ocean system, both in terms of natural drift and oscillations such as ENSO, as well as the response to imposed perturbations like the warming trend due to anthropogenically produced greenhouse gases. In some model simulations, this is leading to increased precipitation and meltwater runoff at high latitudes, enough to affect the thermohaline circulation and even to shut it off altogether, or throw it into reverse, with potentially catastrophic consequences for the Earth and its inhabitants.

There are also issues of the *predictability* of the climate system. It is well known that even short-term weather forecasting is limited by the fact that the atmosphere, even when coupling to the ocean is neglected, is a complex and in many ways (not all of them understood) essentially chaotic system. What, then, are the prospects for climate forecasting, involving the fully coupled system (and possibly variations in forcing due to changes in the solar input as well) over much longer time scales? Studies of large-scale oscillations in the current climate, in particular of ENSO, indicate that some gross changes can be forecast years and decades ahead. Slow trends can also be expected to yield to analysis once we have more detailed and precise numerical models and better global databases.

Perhaps the most interesting, and urgent, problem is that of understanding abrupt changes or step functions in key climate variables. These are theoretically possible in complex systems, and there is evidence of rapid climate change in the Earth's past. These may have been due to changes in the thermohaline circulation as discussed above, but exactly what happened and the range of possibilities for the future remain difficult to model and predict.

In summary, we need a better understanding of

(1) the relationship between atmosphere and ocean dynamics that determines the current climate;
(2) current and future gradual trends leading to global change;
(3) periodic oscillations on a global scale, such as the ENSO;
(4) rapid or abrupt transitions, like those that occurred going into, during, and leaving the ice ages;
(5) the 'noise' (fluctuations of a chaotic nature), in the climate system in general and the ocean in particular.

We will revisit these questions in chapter 11.

## Further reading

*Climate System Modelling*, K.E. Trenberth, Cambridge University Press, Cambridge, 1992.

*Thermodynamics of Atmospheres and Oceans*, J.A. Curry and P.J. Webster, Academic Press, London, 1999.

*Introductory Dynamic Oceanography*, S. Pond and G.L. Pickard, 3rd Edition, Butterworth-Heinemann, Oxford, 1993.

*Atmosphere–Ocean Dynamics*, A.E. Gill, Academic Press, New York, 1982.

*The Atmosphere and Ocean, A Physical Introduction*, N. Wells, Taylor & Francis, London, 1986.

## Questions

1. Estimate the mass of the global ocean, given that its mean depth is 3.8 km, and the total mass of the atmosphere, assuming a mean surface pressure of $1.013 \times 10^5 \, \mathrm{N\,m^{-2}}$. What is the mass of the atmosphere, as a fraction of the mass of the ocean?

2. From the previous answer, calculate the ratio of the heat capacities of the atmosphere and ocean, and estimate how much the temperature of the atmosphere would change if that of the ocean fell by 0.01K, with all of the heat passing to the atmosphere. Why might it be appropriate to consider a smaller heat capacity for the ocean when considering its interaction with the atmosphere, and what would be the approximate value then?

3. Describe in general terms the vertical temperature structure of the global ocean and the physical phenomena that produce it.

4. What experimental methods are used to measure temperature in the ocean, from (a) ships, (b) land, and (c) satellites?

5. Describe in general terms the global circulation of the ocean. What is the role of salinity in this?

6. Using a simple expressions for the rate of change of temperature in the mixed layer of the ocean resulting from the balance between diffusion and advection in the thermocline region, estimate a value for the vertical advection rate $w$ in the steady state given a typical value of the diffusion coefficient $\kappa = 0.02 \, \mathrm{m^2\,s^{-1}}$ and a thermocline depth of 100 m.

7. Show that a wind stress $S^* \, \mathrm{N\,m^{-2}}$ acting on the surface of the ocean will induce a transport $Q \, \mathrm{m^2\,s^{-1}}$ given by $Q = S^*/(\rho 2\omega \sin\varphi)$, where $\rho$ is the density of water and $\varphi$ is latitude. Comment on the direction of the flow, and on the resulting large-scale circulation in an ocean basin. What mass of water is moved by a typical wind stress of $0.1 \, \mathrm{N\,m^{-2}}$ at 40° latitude, where $f = 10^{-4} \, \mathrm{s^{-1}}$?

8. Calculate the relative vorticity of a meridional flow in uniform depth water where $v$ reduces from $1 \, \mathrm{m\,s^{-1}}$ to $0 \, \mathrm{m\,s^{-1}}$ uniformly over 100 km distance towards the east. If this flow is present at the equator how far north could the current travel before the relative vorticity was compensated by increased planetary vorticity?

9. If the wind stress changes from $1 \, \mathrm{N\,m^{-2}}$ westwards to eastwards over a meridional distance of 500 km, estimate the southward Sverdrup velocity for an effective water column thickness of 1 km.

10. Approximately 5 PW are transferred from low to high latitudes by the ocean, to balance the input from the Sun. If all of this energy were transported in the Gulf Stream, estimate its velocity, giving reasons for any assumptions. How does the real situation differ?

# Radiative transfer

6

## 6.1 Introduction

Radiative transfer theory is the basis for

(1) understanding the transfer of energy within the climate system; and
(2) interpreting remote sensing measurements of climate variables made from satellites.

We have already seen (Chapter 2) how the Sun delivers energy to the Earth as ultraviolet, visible, and near-infrared radiation, and how the Earth maintains an overall energy balance by re-emitting an equal amount of energy as long-wave, thermal infrared radiation. We calculated the total amounts of energy, and the wavelength of peak emission, from the Sun and the Earth, using the Stefan-Boltzmann law and the Wien displacement law respectively, and noted that both of these can be derived from Planck's Radiation law, one of the most important formulae in modern Physics.

For those energy balance calculations, it was sufficient to treat the Sun and the Earth as 'billiard balls' of a certain size and temperature. However, thermal infrared radiation is also crucially important for the transfer of energy *within* the climate system, and some more detailed theory is required for understanding what happens to the $\sim 70\%$ of the solar flux reaching the Earth that is absorbed in the atmosphere at various heights, and at the surface. This energy is moved around between atmospheric layers by radiative transfer[61] before finally finding its way out to space. This chapter deals with the basic physics of these key radiative transfer processes.

The second use of radiative transfer theory, and the basis of a major industry, is for making measurements from space of climate system variables and for monitoring the state of the climate, for example, by mapping the temperature of the land and the sea surface, or retrieving vertical temperature and composition profiles. The radiation escaping from the top of the atmosphere into space carries information about the state of the planet that we can analyse using infrared spectrometers and radiometers on satellites. This technique, called remote sensing, is the subject of Chapter 9, but it has its foundations in radiative transfer theory and we will consider those basics here.

[61] Radiation and atmospheric motions (advection of sensible heat) both transfer energy vertically and from place to place in the climate system. Radiation tends to dominate at lower pressures, because the opacity is smaller and photons become more efficient at transferring energy over large distances, so radiative processes dominate in the stratosphere, and dynamical processes in the troposphere.

## 6.2  Black body or cavity radiation

[62] Planck obtained this formula for what he called 'the energy distribution in the normal spectrum of heat radiation' by combining the experimentally-derived laws of Wien and Rayleigh, after Kirchhoff had shown that the emitted radiance depends only on temperature and wavelength, and not in any way on the properties of the material. 'Understanding this remarkable function', Planck said in his Nobel acceptance speech in 1920, 'promised deeper insight into the connection between energy and temperature, which is, in fact, the major problem in thermodynamics and so in all of molecular physics'. Because the constant $h$ in the formula represents the product of energy and time, he described it as the 'elementary quantum of action', and realised it 'must play a fundamental role in physics . . . which seemed to require us to basically revise all our physical thinking. . . . on accepting the continuity of all causative connections'.

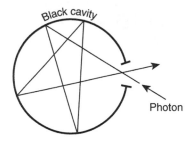

**Fig. 6.1**
A photon entering a cavity through a small aperture has to make many reflections before it has any chance of leaving the cavity. If the aperture is small and the cavity is large, and if the walls of the cavity are blackened and roughened with soot to make them strongly absorbing, the probability of the photon leaving is essentially zero. The aperture is then said to be perfectly 'black', that is, absorbing, and is called a *blackbody*.

[63] Gustav Robert Kirchhoff (1824–1887) is credited with inventing the term 'black body radiation', and with demonstrating that, for a given atom or molecule, the frequencies at which it emits and absorbs are the same. At Berlin University, he taught the young Max Planck, who found him 'dry and monotonous'.

The key equation for calculating the emission of radiant energy is Planck's formula. We used this in Chapter 2, integrated over angle and over wavelength to form the Stefan-Boltzmann law, to calculate the total flux from the Sun and the Earth as a function of temperature. In its basic form, it expresses the *radiance*, that is, the amount of radiant energy leaving unit area of a body of a given temperature per unit time, per unit spectral interval, and per unit solid angle. Radiance is expressed in units of W metre$^{-2}$ (wavelength)$^{-1}$ steradian$^{-1}$ (W m$^{-3}$ sr$^{-1}$), or equivalently in wavenumber units as W m$^{-2}$ (cm$^{-1}$)$^{-1}$ sr$^{-1}$. Through considering the energy density of radiation inside a cavity at a uniform temperature $T$, Planck showed that the radiance $B$ as a function of wavelength $\lambda$ leaving a small hole in the cavity is

$$B(\lambda, T) = \frac{2hc^2}{\lambda^5} \frac{1}{e^{hc/\lambda kT} - 1} \; \mathrm{W\,m^{-3}\,sr^{-1}},$$

where $h$ and $k$ are the Planck and Boltzmann constants and $c$ is the speed of light.

Planck's formula[62] applies to emission from any body, provided it is black. In radiation theory, 'black' has a precise meaning—a black surface is one that absorbs every photon that falls on it. Such a hypothetical perfect absorber, for which absorptance $a_\lambda = 1$ at all wavelengths, is called a *blackbody*. Black-coloured objects in everyday life are not perfectly black, having at least a small reflectivity, but some (a coarse piece of charcoal, for example), come close. Optically thick bodies of gas, like those making up the outer layers of the Sun and the Earth, behave like blackbodies, which is why we were able to apply Planck's and Stefan's formulae to them in Chapter 2, although we noted at the time that this was complicated by the fact that the solar and terrestrial atmospheres are not isothermal but have temperatures that vary with depth, so it is not immediately obvious what value of temperature to use in the formulae.

The nearest thing to a perfect blackbody that exists in the laboratory is a small hole in a large cavity, coated with matt black material inside, like that envisaged by Planck in his original derivation. The reason can be seen by picturing such a cavity as an absorber of radiation (Fig. 6.1). Any photon entering the cavity has to make many reflections (an infinite number if the aperture is negligibly small compared to the radius of the cavity) before it has any chance of leaving, and therefore has virtually 100%, probability of being absorbed ($a_\lambda = 1$ at all $\lambda$). A perfect absorber must also be a perfect emitter (emittance $e_\lambda = 1$ at all $\lambda$), otherwise a black object placed in an isothermal cavity would heat up or cool down to a temperature different from that of the cavity walls. *Kirchhoff's law*[63] makes the more general statement that $e_\lambda = a_\lambda$ is true for all bodies, black or not, and at all wavelengths individually. An easy thought experiment shows that this must also be true, otherwise two cavities joined by a wavelength-selective filter would each heat or cool at the other's expense, contrary to the laws of thermodynamics.

Real bodies, those which exhibit some reflectivity ($r_\lambda \neq 0$), still emit photons according to Planck's formula, but the blackbody radiance it describes must now be multiplied by the emittance $e_\lambda$, so the radiance $R_\lambda$ emitted from a real body is

$$R(\lambda, T) = e_\lambda \, B(\lambda, T).$$

Emissivity[64] can be quite a complex and characteristic function of wavelength, so measurements of $R(\lambda)$ can be used to identify the emitting material, on the Moon, for example, or to study the composition of the atmosphere from an orbiting satellite, as described in Chapter 9.

## 6.3 Atmospheric absorption and emission

Slabs of atmosphere also follow Planck's law, as Fig. 6.2 illustrates. An opaque or 'optically thick' slab of atmospheric gas is shown absorbing radiation with radiance $R$ at wavelength $\lambda$. Because the slab is optically thick, all of the incoming radiation is absorbed and its energy converted to heat. At the same time, the slab at temperature $T$ also emits radiance $B(\lambda, T)$ at the all wavelengths and in all directions.

If radiative equilibrium is assumed, and we neglect other possible sources of heating and cooling of the slab, such as mechanical work done by expansion/compression, conduction from adjacent atmospheric layers, or latent heat release by the condensation or evaporation of water vapour, then the temperature $T$ is determined by the balance between the radiative heating and the emission $B(\lambda, T)$, after both have been integrated over all wavelengths $\lambda$, all surfaces, and in all angular directions.

Thus $\iiint R(\lambda)$ and $\iiint B(\lambda, T)$ are linked through the first law of thermodynamics (conservation of energy), but $R(\lambda)$ and $B(\lambda, T)$ are not otherwise related. $R(\lambda)$ depends on the properties of its source, the Sun at shorter wavelengths and other parts of the atmosphere in the thermal infrared, while $B(\lambda, T)$ is a function only of the slab temperature $T$. In particular, although a slab may absorb and also emit photons at any given

[64] The terms transmission, absorption, emission, reflection, and scattering refer to the respective physical phenomena. When radiation is transmitted, etc. through a particular object (like a layer of atmosphere, or a piece of glass) we refer to the transmittance, emittance, reflectance, and absorptance of the object. When characterising a specific material, we use the terms transmissivity, emissivity, reflectivity, and absorptivity. For example, the transmissivity of quartz at a given wavelength is the transmittance of unit thickness of that material. Somewhat confusingly, the older term *absorption coefficient* is still in common use in place of absorptivity, meaning the absorption per unit amount of absorber, whereas its equivalent, transmission coefficient, has all but vanished completely.

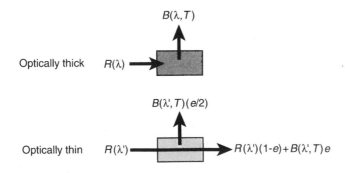

**Fig. 6.2**
Absorption and emission by a slab of atmosphere at temperature $T$ varies with wavelength. If the slab is optically thick (opaque) at wavelength $\lambda$ then the absorptance $a$ (which equals the emittance $e$) tends to unity and the transmitted component vanishes. If at wavelength $\lambda'$ it is optically thin (semitransparent) then there is a transmitted and an emitted component to the emerging radiance. The emitted component depends only on the temperature of the gas and its emittance; the latter depends on the optical path length, as illustrated.

wavelength $\lambda$, in general $R(\lambda) \neq B(\lambda, T)$ since the processes of absorption, thermalization, and re-emission have taken place between the two.

At some other wavelength $\lambda'$ where the same slab is optically thin,[65] only a fraction equal to the absorptance $a$ of $R(\lambda')$ is absorbed, and the rest passes through (Fig. 6.2). The emission in this case is still given by the Planck function, but scaled by the emittance $e$. In the figure, Kirchhoff's law has been used to set $e = a$.

The factor $e/2$ appears in the emission perpendicular to the transmitted beam because the slab in this example is only half as thick in the vertical direction as it is in the horizontal. This makes the point that $a$ and $e$ depend on the number of absorbing or emitting molecules along the line of sight for an optically thin slab. In mathematical terms, the absorptance/emittance of the slab in the long (horizontal) direction is

$$e_H = 1 - e^{-kx},$$

where $k$ is the absorption coefficient and $x$ the physical distance across the slab, so the exponential term is the transmittance. In the shorter (vertical) direction

$$e_V = 1 - e^{-kx/2}$$

and in the optically thick limit

$$kx \to \infty \qquad \text{so } e_V = e_H.$$

whereas for the optically thin case, expanding the exponential as a series,

$$kx \to 0 \quad \text{and} \quad e_V \approx e_H/2.$$

## 6.4 Atmospheric radiative transfer

Figure 6.3 shows a slightly more complicated representation of the behaviour of a single atmospheric layer. It does all of the following:

- absorbs some of the direct solar radiation
- absorbs solar radiation reflected from below (from clouds, or the surface)
- emits thermal radiation at its own characteristic temperature
- absorbs thermal radiation from other layers, above and below
- transmits some of the solar and thermal radiation from all sources
- reflects (by scattering) solar and thermal radiation from all sources, although this is negligible at infrared wavelengths unless the layer contains cloud or aerosol particles.

When we consider this behaviour multiplied by a large stack of layers, enough to represent the whole atmosphere, we realize that computing the amount transfer of energy by radiation can be a complicated business (Fig. 6.4). It can be simplified somewhat by recognizing that solar and thermal radiation calculations can be separated, particularly since the two wavelength regimes do not overlap significantly. In addition, we can often treat clouds in a simplified way. The most basic treatment of clouds assumes

[65] 'Optically thin' is defined as the case where the optical depth $\tau$ of the medium is $\ll 1$, that is, the transmittance $\mathcal{T} = e^{-\tau}$ is close to 1 and both absorptance $a$ and emittance $e$ are small ($\to 0$). 'Optically thick' means $\tau \gg 1$ so $\mathcal{T} \to 0$ and $a$ and $e \to 1$.

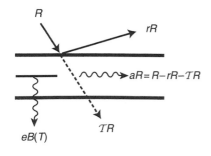

**Fig. 6.3**
Energy is conserved between transmission, absorption and reflection of radiance $R$ at wavelength $\lambda$ by a representative atmospheric layer of reflectance $r$, absorptance $a$, and transmittance $\mathcal{T}$. If the layer is maintained by its surroundings in an isothermal state at temperature $T$, emission also occurs as shown, with emittance $e$, but this is independent of the processes involving $R$.

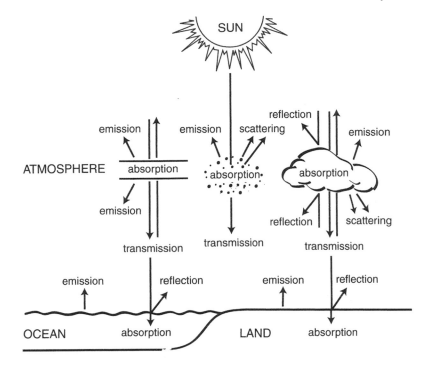

**Fig. 6.4**
Transmission, absorption, emission, and scattering of radiation takes place in each element of the climate system. To build a realistic climate model, we have to decide which of these radiative transfer processes are the most important under a given set of circumstances, and which can be neglected. Then we need fast, accurate schemes for devised for calculating the rate at which energy is transferred within the atmosphere.

that they are characterized by a diffuse reflectivity at solar wavelengths, and switch to behaving like blackbodies at the appropriate atmospheric temperature in the thermal infrared. These are actually quite good assumptions for thick clouds under many realistic conditions.

Dispersed particles (aerosols) found almost everywhere as thin (or not-so-thin) hazes in the atmosphere and even, at short wavelengths, air molecules themselves, are prolific scatters and alter the direction taken by photons that impinge upon them, as well as absorbing and emitting. For this reason, and to deal accurately with clouds, the physics of the scattering process has to be dealt with in advanced calculations. A popular way of doing this is to calculate the scattering properties of a single particle (itself not an easy task, even for the simplest case of a spherical droplet of uniform composition), and then use matrix algebra to compute the reflection, transmission, and absorption of rays travelling through assemblages of particles over a range of representative directions. These matrix operators can then be multiplied to describe the whole atmosphere.

Such complex radiative transfer algorithms are beyond the scope of the present book. We will deal primarily with the simplest case—an atmosphere free of cloud or other scattering particles, with propagation of radiation only in the vertical direction, up and down (a 1-d model). We seek to set up the *radiative transfer equation* (RTE) which describes this behaviour, and which can be solved in various useful cases to obtain the radiance.

## 6.5 The radiative transfer equation

Consider monochromatic radiance $R$ at some wavelength $\lambda$ in the thermal infrared ($\lambda$ subscripts not shown for simplicity) passing through an element

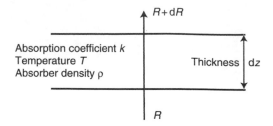

**Fig. 6.5**

In the thermal infrared, the radiance $R$ impinging on a layer of atmosphere is attenuated by absorption and enhanced by emission within the layer, so the emerging radiance $R + dR$ may be larger or smaller.

of absorbing medium (such as a layer of atmosphere) of thickness $dz$, as shown above (Fig. 6.5). The layer will absorb some of the incident radiation, but also augment the beam by emitting at its temperature $T$ according to Planck's formula. The net change in crossing the layer can be written

$$dR = -Rk\rho \, dz + B(T)k\rho \, dz$$

The first term on the right uses the Beer-Lambert law of absorption,[66] which states that the absorptance $a$ is proportional to the product of absorber density and path length, $\rho \, dz$. The constant of proportionality is the absorptivity (also called the absorption coefficient), $k$. The second term is based on the Planck blackbody function $B(T)$ as the source function, and Kirchhoff's law, which tells us that the emittance and the absorptance are identical. $R, k$, and $B$ are frequency dependent, and, in the atmosphere, $\rho$ and $T$ are functions of altitude $z$.

We wish to calculate the radiance $R_\infty$ leaving the top of the atmosphere. Let $-k\rho \, dz = d\tau$, where $\tau$ is a quantity known as the *optical thickness* (or, when measured from the top down, the *optical depth*) of the layer. Then:

$$dR/d\tau = R - B,$$

where $R$ and $B$ are functions of $\tau$. This is *Schwartzchild's equation*, originally developed in an astrophysical context.[67] It can be solved by multiplying both sides by $e^{-\tau}$ to get

$$e^{-\tau}\frac{dR}{d\tau} - e^{-\tau}R = -e^{-\tau}B$$

and integrating from the top of the atmosphere ($z = \infty$, where $\tau = 0$) to the bottom ($z = 0$, where $\tau = \infty$ if the atmosphere is optically thick, that is, if it is deep enough and has a high enough absorption coefficient to be opaque to radiation at the wavelength of interest over its whole depth). Then

$$R_\infty = \int_0^\infty Be^{-\tau} \, d\tau.$$

[66] August Beer (1825–1863) was a professor in Bonn who, in 1854, published a book 'Einleitung in die höhere Optik.' Johann Heinrich Lambert (1728–1777) wrote a book on the passage of light through absorbing media in 1758, while working as a tutor to an aristocratic family in Switzerland. He was later elected to the Prussian Academy of Sciences in Berlin and become a colleague of Euler and Lagrange.

[67] Most of the basis for radiative transfer theory was first developed to understand energy transfer in stellar interiors, in the early part of the twentieth century. It was not until the 1950s that a similar formulation was applied to the atmosphere. Karl Schwarzschild (1873–1916) was an astronomer who worked on variable stars, optical systems, and relativity theory.

The transmittance $\mathcal{T} = e^{-\tau}$ definition, and $d\mathcal{T} = -e^{-\tau}\,d\tau$, so the RTE is

$$R_\infty = \int_0^1 B\,d\mathcal{T} = \int_0^\infty B\,\frac{d\mathcal{T}}{dz}\,dz + e_0 B_0 \mathcal{T}_0$$

The second term represents the emission from the surface with emittance $e_0$ at temperature $T_0$, where $B_0 = B(T_0)$ and $\mathcal{T}_0$ is the transmittance of the whole atmospheric column.

## 6.6 Integration over wavelength

The monochromatic radiative transfer equation tells us how much energy is transferred in a very narrow spectral interval, one over which the transmission varies only slightly so it can be considered to be constant. In general, for energy balance calculations, we need to integrate the RTE over wavelength in order to get the total energy transferred. There is no analytical way to do this, as the opacity of the atmosphere varies rapidly with wavelength due to the various absorption bands. Generally, it is necessary to set up a numerical integration scheme on a sufficiently powerful computer to sum the contributions from a large number of narrow spectral intervals, each of which is the subject of a monochromatic calculation. Occasionally, useful, or at least insightful, results can be obtained by approximations that bypass the need to carry out the full wavelength integration, such as the 'grey' approximation in which wide spectral ranges are represented by a single, mean absorptance value.

Although it can be tedious to perform, the integration of the radiative transfer equation yields powerful results. For instance, when radiance is integrated over all wavelengths and over all solid angles in a hemisphere, it becomes the flux, the total energy per second crossing unit area. If we also integrate up to the top of the atmosphere, we then find the total rate of cooling of the Earth much more precisely than with the simple 'billiard ball' model of the Earth used in Chapter 2. Another use of the spectrally integrated radiative transfer equation is to calculate the vertical temperature profile in the atmosphere that corresponds to radiative equilibrium at every level (see Section 6.8). Satellite observations of the radiance, integrated over narrow spectral bands, can be interpreted in terms of the emitting temperature of the atmospheric layers beneath by using the radiative transfer equation to link the two (Section 6.9). In this way the actual vertical temperature profile can be measured from space, the basis for remote temperature sounding by weather satellites (Chapter 9).

## 6.7 Spectral properties of atmospheric gases

The vibrational and rotational energies of the atmospheric molecules are quantized, so each molecule has a characteristic set of energy levels, separated typically by of the order of one wavenumber for rotational levels and of the order of a thousand wavenumbers for the higher-energy vibrational levels.[68] The position and spacing of these levels can be calculated accurately, although it is beyond our scope to do that here.[69] If the oscillation

[68] The student of Climate Physics is assumed to be familiar with at least the basics of Molecular Spectroscopy theory, and we have scope for only the briefest of summaries here. Those who are not should refer to an introductory text such as Banwell (see under 'Further reading').

[69] Useful results are possible from simple classical models of the molecule, in which the atoms are represented as point masses and the bonds between them as linear springs. For maximum accuracy a complete quantum-mechanical approach of considerable complexity is required (see the texts by Steinfeld and Atkins listed under 'Further reading').

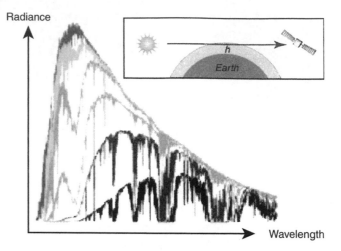

**Fig. 6.6**
Spectra of the Sun viewed from a satellite
along a tangential limb path through of the
atmosphere. The six spectra correspond to
the six different tangent heights above
Earth's surface of the path from Sun to
satellite (height $h$ in the inset) = 90, 40,
30, 20, 10, and 6 km. Note the absorption
in spectral lines and bands of atmospheric
minor constituents.

due to vibration and/or rotation of a molecular dipole moment, or one of its
harmonics, matches that of the photon, absorption (and also emission) can
take place with a high probability.

### 6.7.1 Spectral lines

Photons from the Sun that match the allowed transitions between the energy
levels corresponding to the internal vibrational or rotational modes of
$CO_2$, $H_2O$, $O_3$, and the other 'greenhouse' gases (Section 7.3) are removed
on passage through the atmosphere by absorption. Those that do not have
this match pass through the atmosphere and reach the surface. The energy of
the absorbed photons becomes part of the internal energy of the molecule, to
be released later as emission or 'thermalized' by collisions that convert the
internal energy to kinetic energy of the colliding molecules, thus heating the
atmosphere. The photon stream from the Sun that reaches the ground has
these frequencies attenuated; a spectrometer records narrow bites out of the
solar continuum (which, as we have seen, resembles a 6,000K blackbody)
as shown in Fig. 6.6. These are called spectral 'lines', because they appear
as black lines when the spectrum is dispersed onto a photographic plate,
which used to be the normal way to record a spectrum before the arrival of
electronic detectors.

### 6.7.2 Vibration–rotation bands

Most of the spectral features in Fig. 6.6 are in fact the unresolved groups
of lines that we call bands. Each band corresponds to a single vibrational
transition, with fine structure introduced by dozens of accompanying rotational transitions that add or subtract to the vibrational energy transferred,
producing a family of closely spaced lines that often overlap each other.

The variations of absorptance (and emittance) with wavelength in the
atmosphere depend on the strengths and positions of the spectral lines, how
they are grouped into bands, and where the band centres lie in relation to
the Planck function. Its position in the spectrum determines how effective

a band of a particular species is at absorbing infrared emission from the surface, from other layers in the atmosphere, and from the Sun: a strong band at a wavelength where the emission is low can be less effective than a weak band near the wavelength of peak emission. Similarly, a weak band that coincides with a strong band of the same or another species makes a far smaller contribution than it would if it lay in a part of the spectrum that is otherwise free from absorption.

Molecular spectroscopy theory, which in its simplest form is based on semi-classical models of the vibrating and rotating molecules, allows not only the energy levels, but the transition probabilities between them, to be calculated, giving the positions of the spectral lines and bands.[70] Because each molecule has different vibrational and rotational modes, its energy levels and transitions are unique, so the species giving rise to absorptions in spectra like those in Fig. 6.6 can be identified by the wavelengths at which they absorb and emit. In order to calculate their role in the greenhouse effect, however, we need to know not just where in the spectrum their individual lines and bands act and how strongly, but also how they combine over broad spectral intervals and along inhomogeneous atmospheric paths. This is where radiative transfer theory takes over from molecular spectroscopy.

### 6.7.3  Band absorption formulations and band models

Spectral lines are not restricted to a single sharp wavelength, but have a finite width and a definite shape that depends on the physical conditions, especially temperature and pressure. These can also be calculated from the detailed quantum-mechanical theory of the molecular structure, and they can be measured in the laboratory by passing radiation through a sample of gas of known temperature, pressure, and path length in a cell and analysing the spectrum, measured at high resolution using laser spectrometers. Vast catalogues of line positions, strengths, and widths are available for all common atmospheric gases.

Using these, the transmission in the atmosphere, and, at long wavelengths, the emission, can be calculated using the radiative transfer equation. The integration over wavelength and height is relatively simple to calculate by numerical integration on a fast computer, but for tasks where other uncertainties dominate, including estimating atmospheric greenhouse warming, useful approximate formulations are also available. These include *band models*, in which the positions of the lines in the band are assumed to be randomly (or, in some cases, regularly) spaced, so that an analytical formula can be derived for the absorption coefficient.

Since they were introduced in the 1950s,[71] a large number of band models has been proposed, based on a variety of different assumptions about the statistics of the spectral line properties. Some work better than others for a given band of a particular molecule, and the user generally decides which model to use by testing several against an exact 'line-by-line' calculation first. What all band models have in common is a simple analytical form, so that, once validated, they massively simplify the calculation. In a full greenhouse model calculation, which involves many bands of many species,

[70] The theory also tells us the *strength* of each line, that is, the probability of absorption taking place during an individual photon–molecule collision, and the *width* or spread of the line, an effect due ultimately to the uncertainty principle and to Doppler shifting of the photon frequencies.

[71] Most notably by Richard Goody, A statistical model for water vapor absorption, *Quart. J. Roy. Meteorol. Soc.*, **78**, 165–169, 1952.

many atmospheric levels and all wavelengths they can save literally weeks of computer time, allowing greenhouse calculations to be run on a relatively small computer.

[72] For details of this and other spectral band models see, for example, R.M. Goody and Y. Yung, *Atmospheric Radiation*, Oxford University Press, New York, 1989.

For example, in the *random band model*,[72] a band of randomly distributed lines all with the same width $\alpha$ has a transmittance in a spectral interval $\Delta v$ that can be expressed in terms of the average line strength $S$ and line spacing $\delta$ by

$$\mathcal{T}_{\Delta v} = \exp\left(-\frac{Su}{\delta}\left(1 + \frac{Su}{\pi\alpha}\right)^{-1/2}\right)$$

as a function of the absorber amount in the path, $u$. Expressions like this are relatively simple to evaluate and, although most spectral bands have line distributions that are neither regularly nor truly randomly spaced, the results are remarkably accurate, often to within about 10% of the much more arduous numerical integration method.

We can also use this expression to quantify the difference between strong and weak absorption, having already appreciated that, in general terms, the absorption is strong in the spectral interval $\Delta v$ over the absorbing path $u$ if most of the incident energy is transmitted, and weak if most of it is absorbed. We can see from the random band model expression that, if $Su/\pi\alpha \ll 1$ then

$$\mathcal{T}_{\Delta v} = \exp(-Su/\delta)$$

and the optical thickness of the path is directly proportional to the absorber amount $u$. This is the *weak* condition. If, on the other hand, $Su/\pi\alpha \gg 1$ we see that $\tau = -\log\mathcal{T}_{\Delta v}$ is proportional to the square root of $u$, the so-called *strong* condition. What is happening physically is that the centres of the strongest spectral lines, those that contribute most of the absorption, are completely opaque under strong conditions, so if more gas is added the absorption increases only in the line wings. The lines are then said to be *saturated*.

The importance of strong and weak absorption, for present purposes, is that gases present in trace amounts double their optical depth if the abundance along a path doubles, whereas gases present in larger amounts (like

**Fig. 6.7**
A sketch of absorption coefficient versus wavelength in a strong and a weak spectral line. The strong line is completely absorbed in the centre (saturated), and increasing the absorber amount by $x$ increases the area under the line by only $\sqrt{x}$, compared to $x$ for the unsaturated line.

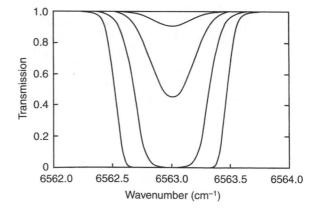

$CO_2$) exhibit absorption that is saturated in the strongest bands, and so have to increase in abundance by four times before their contribution to the optical thickness of the atmospheric path doubles (Fig. 6.7). This square-root dependence of $\tau$ upon $u$ can be verified by writing the random band model expression for $\tau$ in the strong limit:

$$\tau = \frac{Su}{\delta}\left(1 + \frac{Su}{\pi\alpha}\right)^{-1/2} \approx \frac{Su}{\delta}\left(\frac{Su}{\pi\alpha}\right)^{-1/2} = \frac{1}{\delta}(Su\pi\alpha)^{1/2}.$$

For a path with constant pressure or constant mean pressure the only variable on the right-hand side is $u$, which is directly proportional to the mixing ratio $a$ of the absorbing gas, making $\tau$ proportional to $\sqrt{a}$.

## 6.8  Radiative equilibrium models for the temperature profile

Having equipped ourselves with the basics of radiative transfer theory, we now apply it to the first of the main areas of interest cited at the beginning of the chapter—the transfer of energy within the atmosphere. The example we will consider is how we expect the vertical temperature profile to look if it is determined purely by radiative transfer between layers, with each layer in equilibrium with all of the others (including the surface below and space above).

To proceed we note that the integration of the RTE need not be carried out right up to the top of the atmosphere—integration up to some level $z'$ will allow us to calculate the flux into a layer with its base at that height. The downward flux into the top of the layer at $z' + \delta z'$ can be calculated in a similar way. The difference is the flux divergence, $dF/dz$, which results in net heating of the atmosphere according to the expression

$$\frac{\partial T}{\partial t} = -\frac{1}{\rho c_{\mathrm{p}}}\frac{\partial F}{\partial z},$$

where $\rho$ is the air density and $c_{\mathrm{p}}$ its specific heat.

With the assumption that the atmosphere is in radiative equilibrium, that is, neglecting all motions including convection, we can find by an iterative method the temperature profile $T(z)$ that makes $\partial T/\partial t$, according to this equation, equal to zero at every level.

The next step is to solve the full RTE for the flux $F$, which leads to a result like than shown discussed in Chapter 4 (and shown in Fig. 4.3). We can obtain qualitatively similar results using the 'grey' approximation, which bypasses the integration over wavelength by making the assumption that the absorption coefficient at each wavelength can be replaced by a single mean or effective value, which is the same at all wavelengths. This clearly is a sweeping approximation, but often gives remarkably similar answers to a fully detailed calculation and is therefore quite useful. In the extreme form of the approximation, valid when the mean absorption coefficient is large, each layer can simply be treated as a blackbody. With this assumption the

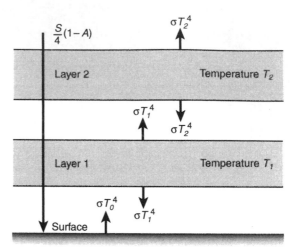

**Fig. 6.8**

A simplified radiative equilibrium model, in which the atmosphere is represented by a series of optically thick layers each at a uniform temperature.

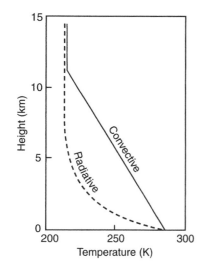

**Fig. 6.9**

Vertical temperature profile calculated assuming radiative equilibrium (dashed line) and radiative-convective equilibrium (solid line). The two converge in the stratosphere, where radiative equilibrium is a good approximation, but in the more opaque troposphere radiation is less efficient and would give a profile that is much steeper than is observed. The convective profile is much closer to reality.

integrated Planck function reverts to the simple Stefan-Boltzmann expression, as we have seen before, and the energy balance calculation reduces to the simple model shown in Fig. 6.8.

Here the surface and two height levels define a two-layer model of the atmosphere (clearly this could be extended to a large number of layers—40 to 100 would be typical—to give a more detailed representation of the real atmosphere). The temperatures at the surface and the two levels bounding the layers are $T_0$, $T_1$, and $T_2$, respectively. Then it is easy to see by inspection of the figure that the layers are in equilibrium when

$$2\sigma T_2^4 = \sigma T_1^4$$

and

$$\sigma T_2^4 + \sigma T_0^4 = 2\sigma T_1^4,$$

while the energy balance equation for the surface is

$$\frac{S}{4}(1-\alpha) + \sigma T_1^4 = \sigma T_0^4.$$

The three simultaneous equations can then be solved to give values for the three temperatures, forming a crude radiative equilibrium temperature profile. If extended to a large number of layers to make the equilibrium profile smooth, the result of such a calculation generates a temperature profile like that shown by the dashed line in Fig. 6.9. However, as we saw in Section 3.5, the real atmosphere shows a profile closer to the solid line in this figure, confirming that the true tropospheric temperature profile is not the result of radiative equilibrium alone, since the radiative equilibrium profile is unstable against convection (super-adiabatic). If the air were as cool at say 5 km altitude as radiative equilibrium predicts, it would rapidly sink and be replaced by warmer air from below. Above 10 km, in the stratosphere, the difference becomes small because the atmosphere is no longer optically thick and radiation becomes more efficient than convection at moving energy between layers. Near the surface, the air is too opaque to infrared radiation for this to be the case.

## 6.9 Temperature sounding and weighting functions

Now we look at the second of the principal applications of radiative transfer theory, the use of radiation measurements to measure climate variables from space. We focus on the example of measuring the vertical temperature profile in the atmosphere, including the surface temperature at the lower boundary. Assuming that the upwelling radiation is being measured by an instrument that is looking vertically downwards, and so viewing the complete atmospheric column, how can we obtain vertically resolved information about the temperature?

The answer, of course, is to select several spectral intervals of different opacity in which to measure the radiance. In some, inside the strong absorption bands of atmospheric gases, the atmosphere will be highly opaque, and the radiation measured must originate near the top, since the instrument cannot 'see' any deeper. At the other extreme, in an atmospheric 'window' (a very transparent spectral region), the instrument will look right through the atmosphere and 'see' only the surface. At wavelengths where the absorption has an intermediate value, the radiance originates somewhere in the middle of the atmosphere.

A radiometer measures the intensity of the thermal emission from the target in each of a set of narrow spectral bands, and this is converted to temperature by assuming Planck's function for the emission from a blackbody as the source function in the radiative transfer equation. When viewing the atmosphere vertically downwards, the emission can be thought of as originating from a series of layers, each of which contributes to the measured radiance (Fig. 6.10). The weighting of these contributions varies with the wavelength at which the measurement is made; layers nearer the top of the atmosphere contribute more in spectral regions where the atmosphere is more strongly absorbing, for example, in a vibration–rotation band of an abundant infrared active species such as carbon dioxide. In each case, the contribution as a function of height is described by a wavelength-dependent *weighting function*.

Consider the RTE as it applies to a narrow interval $\Delta\lambda$ centred on wavelength $\lambda$. The radiance at the top of the atmosphere in this interval is:

$$
\begin{aligned}
R_\infty(\lambda, \Delta\lambda) &= \int_0^\infty B(\lambda, T) \, d\mathcal{T}(z, \lambda, \Delta\lambda) \\
&= \int_0^\infty B(\lambda, T) \frac{d\mathcal{T}(z, \lambda, \Delta\lambda)}{dz} \, dz + B(\lambda, T)\mathcal{T}(0, \lambda, \Delta\lambda),
\end{aligned}
$$

where the Planck function $B$ depends on $\lambda$ but not, or only very weakly, on $\Delta\lambda$, because $B$ is a slowly varying function of wavelength and $\Delta\lambda$ is narrow. The transmittance $\mathcal{T}$, on the other hand, can depend quite strongly

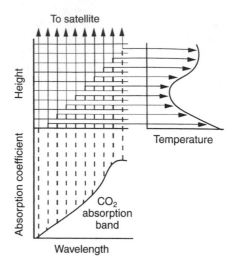

**Fig. 6.10**
The vertical lines represent the different wavelength channels in a temperature-sounding radiometer. If these are positioned in an absorption band, of carbon dioxide say, the radiation detected at the top of the atmosphere corresponds to emission from progressively higher atmospheric layers, as represented by the horizontal lines. These can be used as described in the text to retrieve the vertical temperature profile layer by layer, as shown at top right.

on the exact spectral interval if it is in or near an atmospheric molecular absorption band. It also depends, of course, on the path length from height $z$ to the top of the atmosphere.

If $\Delta\lambda$ is in a strongly absorbing region, $T(0, \lambda, \Delta\lambda)$ tends to zero and

$$R_\infty(\lambda, \Delta\lambda) = \int_0^\infty B(\lambda, T(z)) \frac{dT(z, \lambda, \Delta\lambda)}{dz} \, dz,$$

where we have now included explicitly the dependence of temperature $T$ on height $z$. Inspecting this expression, we see that the radiance leaving the top of the atmosphere is just the weighted mean of the Planck function, and the weighting function is the derivative with respect to height of the transmittance of the atmosphere along a vertical path.

Let us consider what general form the weighting function takes. At the top of the atmosphere, the transmittance $T(\infty, \lambda, \Delta\lambda)$ must be one since the reference point is above all of the absorbing gas. Normally, $T(z, \lambda, \Delta\lambda)$ will make a smooth transition from zero at the surface to one at large $z$, as shown by the dashed line in Fig. 6.11. Its derivative with respect to $z$ then has a definite peak, as shown by the solid line.

It is also intuitively obvious that the whole transmittance curve will move upwards, and along with it the peak in the weighting function $dT/dz$, if the choice of spectral interval $\Delta\lambda$ is changed to one where the atmospheric absorption is stronger, and vice versa. By choosing a series of adjacent spectral intervals along the edge of an atmospheric absorption band, where the absorption coefficient changes rapidly with wavelength, it is possible in principle to select a set of weighting functions that span the atmosphere from top to bottom (Fig. 6.12). The calculation of the weighting functions requires

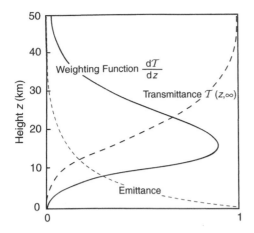

**Fig. 6.11**
Transmittance and emittance profiles and weighting function associated with the radiance leaving the top of the atmosphere in a narrow spectral band.

knowledge of the composition of the atmosphere, at least the abundances of the principal absorbers in the spectral intervals chosen for the sounding instrument. This is one of the reasons why wavelengths dominated by $CO_2$, or in the microwave $O_2$, are chosen in general, since these gases are nearly uniformly mixed.[73]

Since most of the radiation originates near the peak of the relevant weighting function, a set of radiance measurements contains information about the variation with temperature with height. A radiometric instrument on a satellite can make such a set of measurements, which can then be used to reconstruct the temperature profile with a vertical resolution that is related to the number and width of the weighting functions. An optimized set of nadir (downward-viewing) radiances can yield profiles with a temperature accuracy of 1 or 2K with a vertical resolution of around 10 km.

## 6.10 The inverse problem

Solving for the temperature profile, given a set of $N$ radiance measurements in $N$ spectral intervals, each corresponding to a different weighting function, is called the inverse problem, and its solution is called a retrieval. Given that each radiance depends on the temperature profile according to the RTE:

$$R_\lambda = \int_0^\infty B_\lambda(T(z)) \frac{dT_\lambda(z,\infty)}{dz}\,dz + e_\lambda\, B_\lambda(T(0)) T_\lambda(0,\infty)$$

the problem reduces to using $N$ versions of this equation for different values of $\lambda$ to retrieve the temperature profile $T(z)$.

A number of methods have been developed for the purpose. The simplest, which makes the principles involved very clear, is an iterative technique called *relaxation*.[74] This begins by making the simplifying assumption that all of the radiation in a spectral channel comes from the atmospheric level corresponding to the peak of the weighting function (Fig. 6.13).

Then, the radiative transfer equation for the $n$th channel becomes simply

$$R_n = B(T(z_n)),$$

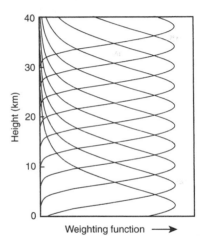

**Fig. 6.12**
A set of idealised weighting functions covering an extensive height range in the atmosphere. For a realistic set corresponding to a real instrument, see Chapter 9 (Fig. 9.9). Weighting functions are usually normalised so that they are dimensionless.

[73] However, if a simultaneous set of observations is made at wavelengths dominated by an absorber of unknown concentration, for example water vapour, then a second retrieval can be performed in which the profile of the absorber amount is the unknown. This is the basis for composition studies by remote sensing, discussed in Section 9.8.3.

[74] M.T. Chahine, A general relaxation method for inverse solution of the full radiative transfer equation, *Journal of the Atmospheric Science*, **29**, 741–747, 1972.

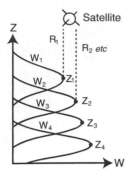

**Fig. 6.13**
Illustrating the inverse problem: $W_n$ is the weighting function for the $n$th spectral channel, which peaks at height $z_n$, and $R_n$ is the corresponding radiance measured at the top of the atmosphere.

[75] This calculation cannot be done analytically, and the usual method is to divide the atmosphere into a large number $L(\sim 100)$ of isothermal layers of equal thickness $\Delta z$, or equal absorber amount $\Delta u$, and replace the integration over height by a summation over layers. Then the RTE takes the form

$$R = \frac{\sum_L B(T(L))[T(L) - T(L-1)]}{\sum_L [T(L) - T(L-1)]},$$

which is readily programmed and solved on a computer.

where $R_n$ is the measured radiance in channel $n$, $T(z_n)$ is the temperature at height $z_n$, and $z_n$ is the pressure at the peak of the weighting function $W_n = \mathrm{d}T_n/\mathrm{d}z$.

Then we can write

$$T(z_n) = B^{-1}(R_n) \quad \text{for } n = 1, 2, \ldots N$$

$$(N = \text{number of measurement channels}),$$

where $B^{-1}$ is the inverse of the Planck function, that is, $T(z_n)$ is the brightness temperature corresponding to the radiance $R_n$ (Section 9.5.3). Thus we obtain a first-order solution for the profile in the form of a series of temperatures at each of the weighting function peaks.

To make use of the information in the shapes of the weighting functions, and to obtain a smooth, continuous profile over the whole range of $z$, next we take the weighted mean of each of these temperatures at every height, thus:

$$T(z') = \frac{W_1(z')T(z_1) + W_2(z')T(z_2) + \cdots + W_N(z')T(z_N)}{\sum_{i=1}^{i=N} W_i}$$

at every $z'$ from 0 to $\infty$, $\infty$ being the top of the atmosphere or, in practice, the height at which the value of the uppermost weighting function falls to a small value relative to its peak.

The next step is to calculate the radiances $R'_n$ from this temperature profile using the radiative transfer equation,[75] and compare them to the measured radiances $R_n$ in each channel. If they are the same within the uncertainty in the measurement, usually determined from the instrumental signal-to-noise ratio (Section 9.5), the solution is complete. Usually, however, it is necessary to follow an iterative procedure until the calculated radiances 'relax' to those that were measured from the satellite. The simplest way to do this is to multiply $T(z_n)$ by the ratio of measured to calculated radiances, $R'_n/R_n$. In doing this we are recognizing that radiance increases monotonically with temperature, so the correction is bound to be in the right direction. By treating this dependence as linear, that is, assuming that radiance is directly proportional to temperature, when we know that the true dependence is a higher-order function, we make sure that we do not over-correct, and that the iteration will converge. In most practical cases, it does so after only a few cycles of the iteration.

Most of the retrieval methods used in practice are based on least-squares fitting of measured to calculated radiances, and often include a number of refinements. For example, instead of finding the profile from the measurements alone, statistical a priori information about the expected value of the profile, based on a history of previously measured values (climatology), can be folded in. One way to do this is to use the climatological value of the temperature profile as the starting point for the iteration in a retrieval by relaxation. In Chapter 9 we will show some profiles of atmospheric temperature and humidity that have been retrieved from radiances measured by operational satellite radiometers, and the design principles of the instruments that are used to gather the radiance data.

## Further reading

*Molecular Quantum Mechanics*, P.W. Atkins and R.S. Friedman, Oxford University Press, Oxford, 1996.

*Molecular Spectroscopy*, J.L. McHale, Prentice Hall, Upper Saddle River, New Jersey, 1999.

*Molecules and Radiation—An Introduction to Modern Molecular Spectroscopy*, Jeffrey I. Steinfeld, MIT Press, Cambridge, Mass., 1985.

*Remote Sounding of Atmospheres,* J.T. Houghton, F.W. Taylor, and C.D. Rodgers, Cambridge University Press, Cambridge, 1984.

*Remote Sensing of the Lower Atmosphere*, G.L. Stephens, Oxford University Press, New York, 1994.

*Atmospheric Radiation: Theoretical Basis*, R.M. Goody and Y.L. Yung, Oxford University Press, New York, 1996.

*An Introduction to Atmospheric Radiation*, K.N. Liou, Academic Press, San Diego, 2002.

*Inverse Methods for Atmospheric Sounding, Theory and Practice*, C.D. Rodgers, Series on Atmospheric, Oceanic and Planetary Physics, Vol. 2, World Scientific, Singapore, 2000.

## Questions

1. Derive the monochromatic radiative transfer equation in the form

$$R_\infty = \int_0^\infty B \frac{\mathrm{d}\mathcal{T}}{\mathrm{d}z}\, \mathrm{d}z + B_0 \mathcal{T}_0,$$

   where $R_\infty$ is the radiance leaving the top of the atmosphere, $\mathcal{T}$ is the transmission to space from height $z$, and $B$ is the Planck blackbody function.

2. Describe in simple terms the random band model for calculating the atmospheric transmission, and the assumptions on which it is based. Obtain simplified expressions for the transmission in the 'strong' and 'weak' absorption limits, and explain what these terms mean.

3. What is meant by the 'grey' approximation, when considering radiative transfer in the atmosphere? Describe how this approximation can be used to develop a simple radiative equilibrium model of the atmosphere by dividing it into two or more layers. Indicate qualitatively how you would expect the result from such a model with many layers to compare the real atmospheric profile in the troposphere and stratosphere.

4. Comment on the significance of the function $\mathrm{d}\mathcal{T}/\mathrm{d}z$ in the answer to Question 1 for remote sounding of the atmospheric temperature profile, and sketch its general form.

5. Show how you could solve for the temperature profile, given a set of radiance measurements in a number of spectral intervals corresponding to different transmission functions. How would you choose the spectral intervals to be used for this purpose?

# 7 Earth's energy budget: The 'greenhouse' effect

## 7.1 Introduction

The presence of the atmosphere raises the mean temperature of the surface of the Earth considerably over what it would be on an airless sphere with the same size, albedo, and distance from the Sun. Ignoring clouds for the moment, the air is largely transparent to the incoming solar flux, but mostly opaque to outgoing thermal infrared radiation. Thus the atmosphere forms a blanket that is colder on the outside than the inside. It presents an effective emitting temperature of about 255K towards outer space, representing equilibrium with the incoming solar power, while the temperature at the base of the atmosphere is around 33K higher.

The key features of the greenhouse effect can be summarized as follows.

1. The energy balance near the surface of the Earth balance is affected principally by the abundance of *minor* constituents in the atmosphere, with the major constituents $O_2$ and $N_2$ playing little or no part.
2. The background or pre-industrial levels of greenhouse gases warm the surface by about 33K and without this 'natural' greenhouse effect the Earth would be frozen everywhere.
3. The abundances of carbon dioxide, methane, ozone, and freons, the most abundant of the 'greenhouse gases' after water vapour, are changing because of human (anthropogenic) activities (e.g. Fig. 7.1, Chapter 10).
4. Simple theory predicts a further surface temperature increase in response to these compositional trends.
5. Complex theory modifies this change due by including positive and negative feedback processes, the net effect of which is not easy to predict.
6. But it still appears, from virtually all models, from the simplest to the most complete and sophisticated, that serious changes in mean surface temperature, of the order of 5K, are in prospect in the present century.

## 7.2 Is the atmosphere a 'Greenhouse'?

It is important to emphasize the difference between the 'natural' greenhouse effect, the existence of which is not in dispute, and which sustains life on Earth, and the possibility of changing it (the *enhanced* greenhouse effect), in particular by burgeoning human industrial activities that may represent a threat to the habitability of the planet, in the form of 'global warming'.[76]

[76] The credit for pointing out the link between potential global warming and human activities is generally credited to Svante Arrhenius (1859–1927). In his paper On the influence of carbonic acid in the air upon the temperature of the ground, *Philosophical Magazine*, **41**, 237–276, 1896, Arrhenius noted that atmospheric carbon dioxide concentrations would increase with the consumption of fossil fuels, and predicted that, if atmospheric carbon dioxide doubled, the Earth would become several degrees warmer. The origin of the term 'greenhouse effect' is obscure, but the first to make the analogy seems to have been Jean Fourier, as long ago as 1827, when he noted that the existence of the atmosphere must raise the temperature of the Earth's surface, and called it 'un effet de verre'.

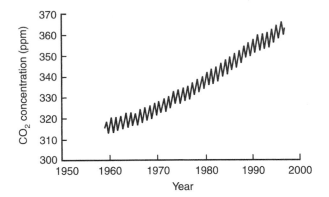

**Fig. 7.1**
The atmospheric $CO_2$ concentrations in parts per million by volume (ppm) at Mauna Loa, Hawaii. The oscillation, superimposed on the secular trend, is the annual cycle caused by seasonal variations in $CO_2$ absorption by plants. (Data from Scripps Institute, University of California.)

The enhanced greenhouse effect is associated in the popular media with controversy, and it is worth considering in this introduction to a discussion of the physics involved why this is so. Once again, no-one disputes the existence of the greenhouse effect, nor its role in maintaining the surface temperature of the planet above the freezing point of water, at a comfortable level for the inhabitants. Neither is there is there any serious dispute over its dependence on the abundances of trace species, so that if carbon dioxide and the others increase enough then an enhancement of the greenhouse is to be expected, with a corresponding increase in mean surface temperature. The argument centres mainly on whether the increase in greenhouse gases already measured, which again is not in doubt (from the evidence in Fig. 7.1, for example), has produced a measurable temperature rise.[77]

The temperature data shows a rise of around 0.6K in the last 100 yrs or so (Fig. 1.12), and we will see later (Section 11.1) that this is close to what models predict in response to the measured composition changes. However, there are at least three reasons why caution should remain. First, the data contain errors, since producing a global average number from a diverse set of data is far from easy. Second, the climate shows natural fluctuations associated with chaotic behaviour, of the kind expected in complex systems, that manifest themselves in slow, cyclical changes in cloud cover, ocean circulation, and so forth. Some of these are apparent in the temperature record as inter-annual and inter-decadal fluctuations that are not much smaller in amplitude than the recent overall warming. It is not possible to say with absolute certainty that the increase in measured global mean temperature (shown in Fig. 1.12, for example) is not due to some internal fluctuation of the climate system with a period of about a century, or longer. Finally, even the best current models have known, and probably some unknown, limitations (inadequate spatial resolution and so forth). Their demonstrated ability to fit the recent run of temperatures, when the answer is known in advance, is a necessary, but not sufficient, reason for having complete faith in their predictions of larger changes in the future.

[77] Some critics of the greenhouse interpretation of global warming would agree that a rise has occurred in global mean temperature, but deny that it can be attributed to human activities, saying the change in atmospheric composition is negligible compared to natural fluctuations or solar variations.

## 7.3  The 'Greenhouse' gases

The magnitude of the greenhouse effect depends on the details of the spectral properties of the gases in the atmosphere, such as the strengths and positions

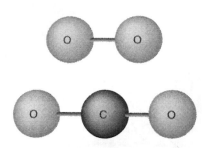

**Fig. 7.2**

Why is carbon dioxide an important 'greenhouse' gas while molecular oxygen is not, despite being many orders of magnitude more abundant in the atmosphere? The reason is the symmetry of the $O_2$ molecule (top), arising from its simplicity, resulting in the absence of an electrical dipole moment to interact with atmospheric radiation. The $CO_2$ molecule (bottom), is also linear and symmetrical but, because it has three atoms, can make asymmetric vibrations, including stretching and bending modes that change the charge separation between C and O atoms and hence the dipole moment.

[78] Except possibly for a change in direction, usually without loss or gain of energy (conservative scattering). Such scattering of photons by molecules, called Raleigh scattering, depends on the fourth power of the wavelength of the photon, and hence is much more important at short (UV) than at long (IR) wavelengths. In the infrared, molecular scattering is negligible compared to that from to aerosol or cloud particles, which of course are larger than air molecules.

of the spectral lines, how they are grouped into bands, and where the band centres lie in relation to the Planck function (Section 2.9). The position of a band of a particular greenhouse species in the spectrum determines how effective that gas is at blocking infrared emission from the surface: a strong band at a wavelength where the emission is low can be less effective than a weak band near the wavelength of peak emission. Similarly, a weak band that coincides with a strong band of the same or another species makes far less additional contribution than it would if it lay in a part of the spectrum that is otherwise free from absorption.

The gas most often cited as primarily responsible for global warming, $CO_2$, forms only about 0.035% of the atmosphere and changes by only about 1 ppm (0.0001% of the total) each year, values that might be thought to render its influence negligible compared to the two principal constituents $N_2$ and $O_2$. That is not the case, however. The fundamental difference between $N_2$ and $O_2$, on one hand, and $CO_2$, on the other, is one of molecular symmetry. All three molecules are symmetric about their centre of mass if the atoms are stationary with respect to one another; but of course they never are. Even in the ground state, the atoms making up the molecules vibrate and rotate with respect to their centre of mass.

In the homopolar diatomics (those with two identical atoms), however, neither rotation nor vibration changes the symmetry. Triatomic molecules like $CO_2$, $H_2O$, and $O_3$ can bend as they vibrate, or vibrate asymmetrically (with the C moving towards one O and away from the other). As a result they present an oscillating dipole moment to interact with the electromagnetic field of an incoming photon, allowing absorption to occur. There are far more interactions between the photons and $N_2$ or $O_2$ molecules, but the photons normally survive these unscathed.[78] $N_2$ and $O_2$ will not respond until a photon with enough energy to stimulate an *electronic* transition comes along; these are all at ultraviolet wavelengths, or shorter.

Some of the principal gases involved in the greenhouse effect, such as water vapour, carbon dioxide, methane, ozone, and nitrous oxide, occur naturally in the atmosphere but have been enhanced (or sometimes, like stratospheric ozone, reduced) by human activities. Others are entirely man-made, such as the chlorofluorocarbons (CFCs) and hydrofluorocarbons (HCFC). More than half, roughly 60%, of the natural greenhouse effect is due to water vapour, with carbon dioxide ($\sim$35%) contributing most of the rest, with methane, nitrous oxide, and ozone making up the remaining $\sim$5% in roughly equal amounts.

Because of its large contribution to the greenhouse, any progressive change in the mean water vapour amount in the atmosphere will affect global warming. Although, of course, most of the atmospheric $H_2O$ is not of anthropogenic origin, a secondary effect of any additional warming by $CO_2$ or other emissions is likely to be an increase in water vapour due to more rapid evaporation from the oceans and the land. This is difficult to quantify; intuitively, we might expect that the relative humidity will stay about the same, so the absolute humidity will increase, but in fact the real behaviour of the atmosphere is probably more complex. This remains a major unresolved issue in climate forecasting models.

Carbon dioxide is the most important anthropogenically produced greenhouse gas for three main reasons: first, although it is only 0.035% of the atmosphere, it is the most abundant; second, as a product of combustion it is increasing significantly as global industrialisation increases, and third, its strong $15\,\mu m$ band lies near the peak of the Planck function for typical atmospheric temperatures, thus maximizing the effect on radiative cooling. Concentrations of carbon dioxide in the atmosphere are now approaching 360 ppm, compared to a fairly steady value of 280 ppm before AD 1700. The increase is primarily due to fossil fuel combustion for transportation, heating, electricity generation, and cooking; accompanied at the same time by changes in the natural systems of vegetation, of which the world's dwindling forests are the prime example. Estimates typically place the relative contributions of these two broad areas in the ratio 65% for increased greenhouse gases and 35% for reduced activity in the biosphere.

There are other important greenhouse gases, in particular methane, nitrous oxide, ozone, and CFCs. These are all changing in amount and all have strong infrared bands. They are rated behind $CO_2$ principally because they are orders of magnitude less abundant. However, this fact also makes them more dangerous; recalling the difference between strong and weak absorption described in the previous section, these rarer gases enjoy a *linear* increase in absorption with amount present, whereas $CO_2$ is already a strong absorber and has only a square-root dependence.

In order to rate the relative importance of different gases, an index called the *global warming potential* has been devised. This is essentially proportional to the increase in temperature at the surface for each extra molecule of the gas that is added to the atmosphere, calculated using models that take all of the above factors into account, plus any 'indirect' effects such as the effect of increasing one gas on the production or destruction of others, especially water vapour and ozone. The results are obviously model-dependent, but provide a useful guide; those in Table 7.1 were derived in an international study from the outputs of many models, integrated over a 100-yr period.

Chlorofluorocarbons are the most effective greenhouse gases per molecule, but have a fairly low and declining concentration, due to the 1987 Montreal Protocol agreed by 46 nations for the global reduction of CFC production and use. Methane, however, has increased by more than 140% since 1750, due to rice cultivation, domestic grazing animals, termites, landfills, coal mining, and oil and gas extraction, accounting for perhaps 15% of the measured increase in the greenhouse effect. The average concentration of nitrous oxide in the atmosphere is increasing at a rate of 0.2–0.3% per year, due to land-use conversion, fossil fuel, and biomass burning, and the use of nitrate and ammonium fertilizers. Ozone has strong infrared bands but its distribution in space and time is so variable that, even with modern observing systems, its role in the enhancement of the greenhouse effect has been difficult to determine with certainty.

**Table 7.1** Global warming potential per molecule (expressed relative to that of carbon dioxide) for those anthropogenically produced gases whose abundance in the atmosphere is increasing and that are expected to contribute significantly to the greenhouse effect over the next 100 yrs. Note the very high values that apply to some of the halofluorocarbons (HFCs) and halons, species that did not exist in the atmosphere until recently

| Gas | Global warming potential |
| --- | --- |
| Carbone dioxide ($CO_2$) | 1 |
| Methane ($CH_4$) | 21 |
| Nitrous oxide ($N_2O$) | 310 |
| HFC-23 | 11,700 |
| HFC-32 | 2,800 |
| HFC-125 | 1,300 |
| HFC-134 | 3,800 |
| HFC-143 | 140 |
| HFC-152 | 2,900 |
| HFC-227 | 2,900 |
| HFC-236 | 6,300 |
| HFC-4310 | 1,300 |
| $CF_4$ | 6,500 |
| $C_2F_6$ | 9,200 |
| $C_4F_{10}$ | 7,000 |
| $C_6F_{14}$ | 7,400 |
| $SF_6$ | 23,900 |

## 7.4 Energy balance calculations

Now we consider some extremely basic models of the greenhouse effect in which the atmospheric opacity is represented very simply using concepts of atmospheric structure, radiative transfer, and energy balance that were developed in earlier chapters. These models give insight into the physics of the greenhouse process, and some quite useful approximate answers. First, let us consider again the energy balance of the Earth as a whole (Fig. 7.3).

Applying the Stefan-Boltzmann law to obtain the total radiant power of the Sun:

$$E_{Sun} = 4\pi R_S^2 \sigma (T_S)^4$$

and since the same amount of energy must cross a sphere, centred on the Sun, at the distance $R_{ES}$ of the Earth's orbit, we also have:

$$E_{Sun} = 4\pi S R_{ES}^2.$$

Equating the two, we find that the solar constant $S$ (the radiant energy falling on unit area at the mean distance of the Earth from the Sun) is given by

$$S = (R_S/R_{ES})^2 \sigma (T_S)^4.$$

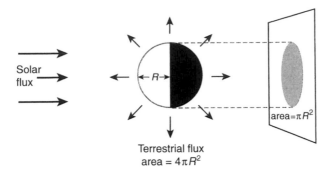

**Fig. 7.3**
The Earth intercepts the incoming solar flux with its cross-sectional area $\pi R^2$, but emits long-wave terrestrial radiation over its entire spherical surface of area $4\pi R^2$.

The total energy radiated from the Earth is:

$$E_E = 4\pi R^2 \sigma (T_E)^4.$$

For equilibrium, this must be equal to the solar energy absorbed by the planet, so allowing for the fraction $A$ that is reflected,

$$E_E = (1 - A)S\pi R^2,$$

from which, with $A = 0.3$, we find that the *effective emitting temperature* (sometimes called the *radiometric temperature*) of the Earth, $T_E$, is approximately 255K or $-18°$C.

The mean *surface* temperature of the Earth, an the other hand, is observed to be about 288K, the difference of 33K being due to the fact that only a small part of the thermal emission to space takes place from the surface: most of it is from the colder parts of the atmosphere, as we saw in Chapter 3. Emission to space actually takes place from all levels in the atmosphere, depending on wavelength, but in an overall (i.e. averaged over all wavelengths) sense we would expect it to peak in the upper part of the troposphere, because these are the highest levels which are optically thick (again in a wavelength averaged sense), obscuring those below, and transmitting through the transparent layers above.[79]

## 7.5 A simple physical greenhouse model

The next step is to extend the energy balance model so that the surface temperature and the effective emitting temperature can be calculated separately. Consider now the surface of the Earth, overlaid by the simplest possible model of the atmosphere in the form of a single homogeneous layer of gas of temperature $T_a$ (Fig. 7.4). To first order the gases of the atmosphere are transparent at those wavelengths corresponding to most of the incoming solar energy, and opaque at those wavelengths at which the Earth emits most of its thermal energy (cf. Fig. 2.10). This behaviour can be approximated by assigning to the model atmosphere a bimodal transmission function with $\mathcal{T} = 1$ at wavelengths shorter than about 4 μm, and $\mathcal{T} = 0$ at longer wavelengths. (This is an example of the grey approximation discussed in Section 6.6).

[79] A detailed calculation shows that the peak emission comes from a level at a height of about 6 to 7 km above the surface, with smaller contributions from below, gradually declining down to the surface, and above, again declining but not reaching zero until around the mesopause. The low-altitude wing represents the contributions in the more transparent parts of the spectrum, including the surface contribution in the 'windows', while the high-altitude wing comes mainly from the strong bands of carbon dioxide and ozone.

**Fig. 7.4**

A 'single-slab atmosphere' model of the greenhouse effect, in which the atmosphere is treated as a homogeneous layer of temperature $T_a$, which is perfectly transparent to solar radiation and perfectly opaque in the thermal infrared. The surface receives the equivalent of two solar constants, raising its mean temperature from 255 to 303K.

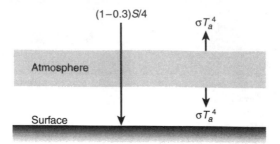

For this situation, the flux of energy $F$ falling on the surface is still equal to the solar constant $S$, multiplied by 0.7 to allow for the albedo $A = 0.3$, and by the factor 1/4 that arises from the spherical geometry. Since $\mathcal{T} = 1$, this falls entirely on the surface, raising its temperature in the absence of further energy input to about 255K as before. The outgoing energy to space, which in this model must be entirely from the atmosphere, since its opacity prevents the surface from radiating to space, must also be equal to the input in order to achieve energy balance. However, the flux from the atmosphere occurs in both the upward and downward directions (Fig. 7.4), and so in equilibrium the surface receives a second contribution equal to that from the Sun, raising its temperature to $2^{1/4} \times 255$ or 303K. This calculated greenhouse enhancement of 48K is rather larger than the observed 33K, not surprisingly in view of the simplicity of the model.

If the atmosphere were homogeneous in reality, this enhancement would be the upper limit on what could be achieved. In fact, it is clear (from studying Venus, for example) that the greenhouse enhancement can be not one but many times the direct solar input.[80] This becomes possible if we consider more realistic models in which the lower atmosphere, still opaque in the infrared, is warmer in its lower regions (which warm the surface) than its upper (which radiate to space). This means representing the atmosphere by two or more slabs or layers, rather than just one. If a large number of isothermal layers is specified, the temperature profile becomes quasi-continuous and a very realistic representation can be achieved.

[80] The effective emitting temperature of Venus is about 240K, while the surface temperature is 730K. The downward flux at the surface must therefore be of the order of $(730/240)^4$ or about 85 times that at the top of the atmosphere.

## 7.6  A better greenhouse model

### 7.6.1  The radiative-dynamical equilibrium profile revisited

A more sophisticated, but still quite basic, model can be formulated by introducing a more accurate representation of the vertical structure of the atmosphere, and by allowing for the spectral band structure in its opacity. The starting point for this is the discussion in Section 3.5 of how the adiabatic (convective) lapse rate in the troposphere, and the radiative equilibrium temperature of the stratosphere, can be obtained from elementary expressions obtained from first principles.

These expressions provide the basis for a simple model of the greenhouse effect which can be used to produce forecasts, albeit naïve ones, of global change arising from an increased atmospheric $CO_2$ burden, and

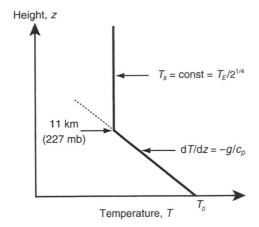

Height, $z$

$T_s = \text{const} = T_E/2^{1/4}$

11 km
(227 mb)

$dT/dz = -g/c_p$

Temperature, $T$

$T_0$

**Fig. 7.5**
A simple one-dimensional model for the
vertical temperature profile, based on the
formulae derived in Chapter 2, and
assuming a tropopause height of 11 km
(227 mb).[81]

[81] This value for the tropopause comes
from the 'US Standard Atmosphere', a
reference compilation of atmospheric parameters, and represents a global mean value.
Since $dT/dz$ is known, we could equally
have adopted the USSA standard value for
the surface temperature (288.15K) to define
the tropospheric profile.

to demonstrate the role of key feedback processes, for example, those
involving clouds.

The model atmosphere, shown in Fig. 7.5, is completely characterised
by a fixed tropospheric lapse rate $\Gamma = g/c_p$, the stratospheric temperature
$T_s$, and either the tropopause height $z_t$ or the surface temperature $T_0$. The
independent variables are the amount of absorber present in the atmosphere,
actually the stratosphere, since the troposphere is assumed to be opaque, and
the value for the mean albedo or reflectivity $A$ of the Earth. The absorption
is represented by a single species with mixing ratio by volume $a$ (which
can be assumed all to be $CO_2$, a fair approximation in the infrared for the
stratosphere, where the water vapour abundance is low).

Since the lapse rate of the troposphere and the temperature of the lower
stratosphere are both constant, the 'climate' is represented graphically as a
temperature profile of two straight lines with fixed gradients ($dT/dz = -\Gamma$
and zero, respectively). The transition between the two, the tropopause,
is at the level $z_t$ where the optical depth, measured from the top of the
atmosphere down, makes the transition from being optically thin, where
radiation is the most efficient means of transferring energy vertically, to
becoming optically thick, where convection takes over and the adiabatic
lapse rate applies. Most of the cooling to space by the atmosphere takes
place from the region just below the tropopause, since the stratosphere,
being optically thin, cannot be a strong emitter (recalling Kirchhoff's law,
section 6.2), and the deep troposphere, near the surface, cannot emit much
radiation directly to space because of the large overlying opacity. The peak
emission occurs from layers at temperatures of around 255K, which, as we
saw in Section 2.9, is the effective radiometric temperature of the Earth
viewed from space.[82]

This transitional optical depth is expected to be invariant, so if the
absorber amount in the stratosphere increases, the tropopause must move
to a lower pressure to keep the same overlying optical depth. Similarly, if
the troposphere gets warmer, due, for example, to a fall in albedo allowing
more solar heating, or an increase in the solar constant, the stratosphere must
also get warmer in order to prevent the tropopause from rising to a lower

[82] Again, we must keep in mind that 'optically thick', 'optically thin', and 'effective
radiometric temperature', are being used
here in a wavelength-averaged sense. In
a few spectral 'windows' like that near
12 μm, the troposphere is transparent right
down to the surface, allowing some cooling
directly to space. Similarly, the stratosphere
is optically thick in a few strong bands,
most notably the 15 μm band of $CO_2$, over
the whole of its height and even up into
the mesosphere, so some cooling to space
occurs from these.

**Table 7.2** Constants used to calculate model temperature profiles, and the radiative equilibrium, stratospheric and surface temperatures that result for present-day $CO_2$ abundances

| | | | |
|---|---|---|---|
| Distance, Earth to Sun | $D_{ES}$ | 1,50,000 000 | km |
| Radius of Sun | $R_S$ | 6,96,000 | km |
| Radius of Earth | $R_E$ | 6 380 | km |
| Effective temperature of Sun | $T_S$ | 5,780 | K |
| Solar constant | $S$ | 1,368 | W m$^{-2}$ |
| Albedo of Earth | $A$ | 0.30 | |
| Effective temperature of Earth | $T_E$ | 255 | K |
| Temperature of stratosphere | $T_s$ | 215 | K |
| Temperature of surface | $T_0$ | 288 | K |
| Pressure at tropopause | $p_t$ | 227 | mbar |
| Tropospheric lapse rate | $\Gamma$ | 6.6 | K km$^{-1}$ |
| Scale height | $H$ | 8.45 (Surface) | km |
| | | 6.35 (Tropopause) | |

pressure. We recall from Section 3.4.2 that the stratospheric temperature $T_x$ in a model that treats the region as an optically thin layer of gas of emittance and absorptance both equal to $e(e \ll 1)$, is given by

$$e\sigma(T_E)^4 = 2e\sigma(T_s)^4$$

so $T_s$ is related to the effective radiative temperature of the Earth $T_E$ by

$$T_s = \frac{T_E}{2^{1/4}} = \frac{255}{2^{1/4}} \approx 215K.$$

The initial model temperature profile is based on the values listed in Table 7.2, which are representative means of the present-day atmosphere.

### 7.6.2  Surface temperature versus absorber amount

Now consider the effect of changing the absorber amount, including the case where the amount of $CO_2$ is doubled, a standard case for climate modellers (the resulting change in $T_0$ is sometimes called the *climate sensitivity*) and a situation which is expected to occur on Earth (for $CO_2$ at least) during the next fifty or so years. The amount of absorber in the isothermal column above the tropopause at pressure $p_t$ is $u = Hap_t$, where $a$ is the mixing ratio by volume of the absorbing gas and $H$ is the atmospheric scale height, both assumed constant with height, a reasonable approximation for the lower stratosphere, where $CO_2$ dominates the infrared absorption. The optical depth in the $CO_2$ bands above the tropopause is proportional to the square root of the product of the line width $a$, which is proportional to pressure, and the absorber amount $u$.[83] Since the optical depth above the tropopause is fixed, $p_t$ must be proportional to $1/\sqrt{a}$ and will move to $1/\sqrt{2}$ its present value if $a$ doubles, as indicated in Fig. 7.6. Since the troposphere lapse rate $g/c_p$ must remain the same, there has to be an increase $\Delta T$ in the surface temperature in the model.

[83] These dependencies were derived in section 6.7.3 for the strong limit of the random band model, which has been shown to give good results for carbon dioxide bands in the atmosphere. It is not necessarily inconsistent to assume the strongly absorbing limit for the $CO_2$ bands, while at the same time assuming that the path for radiative cooling from the tropopause to space is optically thin, as we did when setting up the model; the path in the latter case involves the whole of the thermal infrared spectrum, which is optically thin on the average, despite containing a few strong $CO_2$ (and $O_3$) bands.

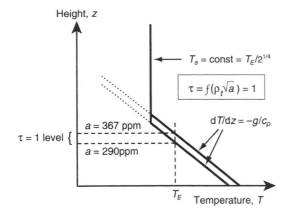

**Fig. 7.6**
Effect of changing the $CO_2$ mixing ratio
throughout the atmosphere on the model
temperature profile. The level at which the
optical depth of the overlying atmosphere
is unity, which is approximately the level
at which the temperature is $T_E$, increases in
height as the proportion of $CO_2$ increases.
Since the lapse rate is (virtually)
unchanged, this raises the surface
temperature

From the expression for pressure as a function of height (see Section 3.4)
we have

$$\Delta p = -\frac{p}{H}\Delta z$$

and since

$$\frac{\Delta p}{p} = \frac{1}{\sqrt{2}} = 0.7$$

and the scale height $H$ at the tropopause is about 4.5 km, the effect of increasing the absorber concentration is to raise the height of the tropopause in the
model by about 3 km. Physically, what is happening is to increase the depth
of the convective layer to offset the greater opacity of the atmosphere, so
that heat is still brought up to the level where it can radiate to space. The
stratospheric temperature remains unchanged because, in the optically thin
approximation, this depends only on the equilibrium temperature of the
planet, $T_E$, which is not a function of $CO_2$ amount (Fig. 7.6).

For a lapse rate of $6\,\text{K km}^{-1}$, the increase in tropopause height that is
predicted for a doubling of $CO_2$ is 18K. By a similar calculation, the effect
of increasing $CO_2$ from its pre-industrial value of 290 ppmv to its current
level of about 367 ppmv is an increase in surface temperature of about 6K.
This result can be used to check how well this simple model performs, since
we have data on the actual global warming that occurred in response to this
change, from compilations of surface temperature measurements made over
the last century (Fig. 1.12).

There is a large amount of scatter in the data, due in part to the natural
variability of the climate system, but also the difficulties inherent in forming
such an average from disparate data. From this plot, we deduce that global
average temperature has increased since pre-industrial times by roughly 1K.
This can be compared to the model increase of about 6K, six times too large.
The prediction of an 18K increase as a result of doubling $CO_2$ is also too high
by about the same factor, relative to predictions from more sophisticated
models, such as those reviewed by the Intergovernmental Panel on Climate
Change (IPCC) that we will discuss in Chapter 11.

Part of the discrepancy is due to the difference between the 'committed' temperature increase (which we have calculated) and the 'realized' temperature change (which we measure), mainly because of the lag introduced by the thermal inertia of the oceans (Section 5.1). The IPCC has estimated that the difference between committed and realized global mean temperature change is 50% if the former is 4.5K and 20% if it is 1.5K. If this is correct, then we can compare half of the predicted increase, that is, 3K, to the observed $\sim$1K, finding there is still a discrepancy of a factor of three.

A further part of this is due simply to the crudity of the model. The 'optically thin, energy balance' approximation for the temperature of the stratosphere is not very exact, especially when considering large changes in $CO_2$ concentration. More exact calculations of $T_s$, as well as measurements from satellites, show clearly that the stratosphere actually cools when its $CO_2$ content is increased, mainly due to the increased emission of radiation to space from the strong 15 $\mu$m band. In the model, this has the effect of raising the height of the tropopause *without* increasing the surface temperature, and so it tends to reduce the surface warming calculated above.

Finally, there is another possibility that should be considered. Perhaps, during the hundred years or so that $CO_2$ and the other greenhouse gases have been increasing, something else has been changing at the same time. The strongest candidate is the albedo of the Earth, which depends on factors like snow and ice cover and cloud abundance, type and height, all things that are very hard to monitor and evaluate. It also depends on the sulphate aerosol loading of the stratosphere, which has also been increasing due to anthropogenic forcing. Let us see what effect changing the albedo has on surface temperature, using the same simple model.

### 7.6.3   Effect of changing albedo on surface temperature

Now we keep the composition of the atmosphere constant and investigate the effect of changing the albedo of the planet on the surface temperature in the model. Such a change might result if the average amount of cloud in the atmosphere changed for some reason; for example, an increase in the number of soluble airborne particles available to act as cloud condensation nuclei, or in the number of small airborne particles (aerosols) of all types dispersed throughout the atmosphere as fine hazes.

Although we do not have enough reliable data to say for sure, it seems entirely possible that such a change accompanied the build-up of $CO_2$ following the industrial revolution, and likely that the aerosol and cloud build-up (if any) would increase rather than decrease the reflectivity of the planet (although not invariably, since some types of cloud and aerosols are good absorbers and can add a net enhancement to the greenhouse effect, see Section 4.10).

Examining the model equations, we see that only the expression for the temperature of the stratosphere depends on the value of $A$. We have

$$T_s = \frac{T_E}{2^{1/4}} = \left(\frac{(1-A)S}{4\sigma}\right)^{1/4}.$$

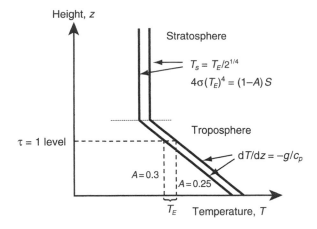

**Fig. 7.7**
Effect of changing Earth's albedo, $A$, from 0.3 to 0.25, on the model temperature profile. The effective temperature of the Earth, $T_E$, falls with increasing $A$, so the stratospheric temperature and the temperature at the $\tau = 1$ level must fall also, dragging down the surface temperature with them.

**Table 7.3** Model results for surface temperature corresponding to different values of Earth's albedo $A$, and different atmospheric $CO_2$ mixing ratios $a$

| Surface temperature (K) | $a = 290$ ppmv | $a = 367$ ppmv | $a = 580$ ppmv |
|---|---|---|---|
| $A = 0.25$ | 288 | 294 | 306 |
| $A = 0.3$ | 284 | 290 | 302 |
| $A = 0.35$ | 280 | 287 | 299 |

The effect of increasing the albedo is to reduce the effective temperature of the planet, and therefore that of the stratosphere. A fall in stratospheric temperature must drag down the surface temperature, since neither the tropopause height nor the lapse rate will change, if the absorber amount stays the same (Fig. 7.7).

Table 7.3 shows some actual values, from which we can see that quite reasonable changes in the net albedo from 0.25 to 0.35 (similar to the range of uncertainties in our ability to determine the actual value of $A$ from measurements) can, in this model, alter the surface temperature by more than 6K.

We could, therefore, completely reconcile the model with observations by assuming that the approximately 1K of warming observed in the last 100 yrs is due to the observed increase in $CO_2$ ($\sim$3K after allowing for the 'hysteresis' in the climate system), offset by an increase in the albedo from 0.275 to 0.3, say, due to pollution-induced changes in clouds and aerosols, which induces a cooling of $\sim$2K, leaving a net rise of $\sim$1K.

In noting this possibility, we recognize of course that this model is too crude to tell us with certainty what is really happening in the extremely complicated system that it represents. It can only give us a very general idea of how the climate *might* respond to known changes, and a feeling for how those responses might reinforce each other, or alternatively tend to cancel each other out.

The next step is to recognize that possible changes are linked to each other, not only in the sense that an increase in aerosol is expected along with an increase in $CO_2$, because of the nature of the industrial processes involved, but because certain changes will *force* other quantities to change, with further consequences (good and bad) for the climate. An example of this we considered earlier is where $CO_2$ increases induce higher surface temperatures, which result in increased evaporation of water from land and sea, which in turn results in higher atmospheric humidities and more rapid cloud formation. Assuming that more cloud means a higher planetary albedo, this is an example of *negative feedback* in the climate system.

Of course, water vapour is itself a powerful greenhouse gas. So, in the above scenario, the increased humidity of the atmosphere is tending to amplify any warming through adding water vapour to the atmosphere, at the same time as it might be reducing it through increased albedo due to clouds. The water vapour enhancement is an example of *positive feedback*.

Another example of positive feedback would be the reduction in the solubility of carbon dioxide in water with temperature. We have already seen in Section 1.4.3 that a large fraction (about one-third) of the $CO_2$ released into the atmosphere is taken up by the ocean and does not contribute directly to global warming. As the ocean gets warmer, however, and as it gets saturated with $CO_2$, it will tend to respond by adding, rather than removing, $CO_2$ from the atmospheric greenhouse, thus amplifying the warming.

To be useful as predictors of future climate change, models have to include these and other feedback processes in their computational schemes. This is perhaps the largest single reason why advanced climate models are so complicated and expensive to design and run. We can demonstrate feedback at work in another simple model, however.

## 7.7  A simple model with feedback

We have seen that, in order to obtain more detailed and more accurate predictions of future changes, climate models need to include feedback processes. The two that have been identified as the most important are the increased humidity of the air that would follow an increase in mean surface temperature, and changes, probably increases, in the amounts of various types of clouds.

Although water vapour is the single most important absorber of those which produce the greenhouse effect, in our simple model, the effect of increasing water vapour is negligible provided the increase is confined to the troposphere. The physical reason is that the lower atmosphere is already optically thick, and the vertical transfer of energy is dominated not by radiation but by convection. This is represented by the fixed lapse rate $\Gamma$ in our model.

Stratospheric water vapour is another matter, but to first order this will not depend at all on the surface temperature because of the cold-trapping effect on water of the low temperature at the tropopause. Carbon dioxide, on the other hand, is approximately uniformly mixed up to heights of around 60 km at least, so an increase in the mixing ratio raises the model tropopause height in the manner described above, but increasing tropospheric water makes no difference at all. (In reality, increased tropospheric water is important because the real atmosphere is not grey, and $H_2O$ is an important absorber in the window regions).

A second feedback process, this time probably tending to oppose any increase in the surface temperature, is the increase in the albedo of the Earth which results from an increase in cloudiness, itself postulated to result from increased humidity and convection in the troposphere. Clouds are generally more reflective than the surface they conceal, so increasing cloud cover tends to increase the albedo of the planet.

Let us assume that, for small changes, the albedo of the Earth depends linearly on the surface temperature, that is, let $dA/dT_0 =$ constant, where the constant is assumed to be positive. With this parameterization included we find that the surface temperature increase of 3K, which results from a 77 ppm increase in $CO_2$ (after invoking thermal inertia effects) is automatically reduced to the observed value of 1K, if an albedo increase of around 0.05 is invoked (Fig. 7.8).

In the $CO_2$-doubling scenario, the cloud feedback effect would have to increase the Earth's albedo to around 0.45 in order to bring the predicted increase of $\sim$9K down until it was in line with results from complex models. Such a large change in the Earth's cloud cover might be easy to detect in the future, but obviously we have no way to know whether the smaller change from $\sim$0.25 to 0.3, which this model would require to have taken place over the last 100 yrs to match current observations, actually occurred, in whole or in part. Even that part which might have taken place since the advent of Earth satellites could easily have escaped detection because of the difficulty in obtaining an accurate global mean measurement.

It is not straightforward even to define what is meant by mean cloud cover in the present sense, since albedo depends on quantities such as cloud height, particle size, geographical location, and so forth. Recently, satellite experiments have been developed which measure the albedo and total thermal emission of the Earth from space, and these will shed light on the actual, as opposed to the conjectured, role of albedo changes in global climate change. However, they will require long series of data covering perhaps as many as 10 yrs before the problem of separating trends from slow fluctuations can be surmounted.

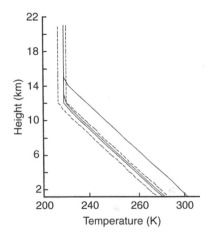

**Fig. 7.8**

A set of models of the atmospheric temperature profile, calculated for different values of $A$, the mean albedo of the Earth, and $a_{CO_2}$, the mixing ratio of $CO_2$ in ppmv. The solid lines have A fixed at 0.3 and, from left to right, $a_{CO_2} = 290, 367,$ and 580, corresponding to pre-industrial, present-day, and possible future Earth. The dashed curves have $a_{CO_2}$ fixed at 367 and, from left to right, $A = 0.35$ and 0.25.

## 7.8  Conclusions

Simple models give some insight into the principal processes at work, and allow us to begin to account for recently observed small changes and to predict major changes in the future. They also show, quite convincingly, that the basic physics implies non-trivial changes in mean surface temperature and/or global cloud cover in response to a doubling of the atmospheric $CO_2$ concentration or other significant changes in greenhouse gas concentration. Finally, they illustrate the principles and main features of the feedback processes that make the final outcome of any global change scenario so uncertain.

What simple models cannot do is to give us a reliable quantitative prediction of the likely changes the Earth will experience in response to the expected ongoing increases in greenhouse forcing. For that we require much more sophisticated models, with dynamical schemes based on coupled ocean-atmosphere GCMs (Section 1.10) and a realistic treatment of radiative transfer as well. Such 'state-of-the-art' models, their predictions and pitfalls, will be discussed in Chapter 11.

## Further reading

*The Greenhouse Effect and Climatic Change*, F.W. Taylor, Reports on Progress in Physics, **54**(6), 881–918, 1991, and *The Greenhouse Effect Revisited*. F.W. Taylor, ibid., **65**, 1–25, 2002.

*Global Warming: The Complete Briefing*, J.T. Houghton, 2nd edition, Cambridge University Press, Cambridge, UK, New York, 1997.

*Climate Change 2001: The Scientific Basis*, Contribution of Working Group I to the Third Assessment Report of the Intergovernmental Panel on Climate Change, edited by J.T. Houghton, Y. Ding, D.J. Griggs, M. Noguer, P.J. van der Linden, X. Dai, K. Maskell, and C.A. Johnson. Cambridge University Press, Cambridge, UK, New York, 2001.

*Atmospheric Radiation*, R.M. Goody and Y.L. Yung, Oxford University Press, New York, 1989.

## Questions

1. Explain the mechanism of the 'greenhouse' effect and how it produces warming of the surface of the Earth. Which gases are 'greenhouse gases', and why are the others, including the two principal constituents of the atmosphere, $N_2$ and $O_2$, not? Would you expect Argon (Ar) and Freon-12 ($CCL_2F_2$) to contribute to the greenhouse effect?
2. Describe a simple model for the vertical temperature structure of the atmosphere in which the stratospheric temperature and the tropospheric lapse rate are both constant.
3. By assigning typical values to the parameters of this model, calculate the surface temperature when the Earth's albedo is 0.30 and the tropopause pressure is 200 hPa.
4. Estimate the change in model surface temperature if (a) the carbon dioxide concentration in the atmosphere doubles, (b) the albedo of the Earth increases from 0.30 to 0.35.
5. Describe a scenario in which the changes (a) and (b) above might actually occur, and how they might be related to each other.
6. Estimate the rate at which the temperature of the oceans would rise if they absorbed all of the energy from the Sun and there was no cooling. What is the corresponding number if you assume that the whole Earth absorbs the heat from the Sun, without cooling?
7. In a simple one-dimensional 'greenhouse' model, the atmosphere is represented by a single layer at the effective radiometric temperature of the Earth. This layer is completely transparent to solar radiation and completely opaque to longer-wave infrared emission from the surface. What is the equilibrium temperature of the surface according to this model, assuming that the radiometric temperature of the Earth is 255K?
8. What does the one-layer model of the previous question predict for the surface temperature of Venus, where the solar constant is twice that at the Earth, and the albedo is 0.76? Why does the answer differ from the observed temperature of 730 K?

# The ozone layer

8

## 8.1 Introduction

Ozone is present near the surface in amounts that are typically between 0.03 and 0.05 parts per million (ppm) in remote areas and somewhat higher, but still less than 1 ppm, in cities. Thus, tropospheric ozone is scarce compared to carbon dioxide (about 350 ppm), water vapour (10,000 ppm or more), and methane (2 ppm). Even in the stratosphere near 25 km altitude, where the ozone concentration peaks, it still reaches only about 10 ppm.

Nevertheless, it is still one of the more important minor constituents of the atmosphere. One reason is because ozone has more strong absorption bands in the ultraviolet (UV) than the other common atmospheric gases, and so it dominates radiative transfer in this part of the spectrum. It also has important bands in the infrared.

Ozone is extremely chemically active, as a result of which it plays a key role in determining the amounts of other constituents with which it reacts. It follows that ozone will be fairly short-lived; in the stratosphere, the lifetime of a typical atmospheric ozone molecule is just marginally longer than the diurnal cycle, at 1 to 2 days. This, plus the fact that it is continually produced on the dayside of the planet by the action of sunlight on ordinary diatomic oxygen, leads to a rather complex and variable distribution, characterized everywhere by a strong peak in the middle stratosphere that we know as the ozone layer.

Thus, the importance of ozone to climate physics is fourfold:

1. The ozone layer acts as a UV filter, removing energetic UV photons in the incoming sunlight before they can reach the troposphere and the surface. The dangers to plants, animals, and humans of over-exposure to such radiation is well known, in particular from photons in the wavelength range 0.28–0.32 μm, known as UVB (see below), which damage DNA. Even with the ozone layer intact, excessive sunburn caused by the few percent of the UVB falling on the top of the atmosphere that reaches the ground is a frequent cause of harm.
2. The absorbed UV energy results in significant heating in the middle atmosphere and alters the mean vertical temperature profile in that region dramatically.
3. Ozone has a number of infrared bands, the strongest of which is at 9.3 μm, near the peak of the terrestrial Planck function, so it contributes to the greenhouse effect (although it makes less than 10% of the contribution of $CO_2$).

4. Ozone is involved in vigorous chemical cycles in the stratosphere and in the troposphere. The former is mainly of interest for understanding the formation and stability of the ozone layer itself; the latter cleanses the troposphere of dangerous organic poisons.[84]

The role of ozone in the greenhouse effect was covered in Chapter 7; this chapter is about the processes that create and destroy ozone, determining its global distribution, and changes in this due to anthropogenic pollution. A detailed treatment of the role of ozone in protecting the biosphere is beyond the scope of this book, except to note its importance as motivation for understanding the physics of ozone behaviour in the atmosphere.

## 8.2   Ultraviolet radiation in the atmosphere

Ozone is produced in the atmosphere by of the action of UV radiation on ordinary (diatomic) oxygen. Only UV photons have enough energy to break the strong bond in the $O_2$ molecule and initiate the photochemical sequences that produce, and destroy, ozone.[85]

About 15% of the solar flux is in the UV part of the spectrum. Three distinct wavelength regions can be defined within the UV by their physical effects:

1. The most energetic photons, with wavelength $\lambda < 0.1 \, \mu m$, can dissociate and/or ionize most atmospheric molecules. The result is the ionosphere, a population of atomic and molecular ions, plus free electrons, mixed in with the remaining neutrals species in the region above the mesopause.[86]
2. The intermediate UV range from $0.1-0.2 \, \mu m$. These photons are mainly absorbed by oxygen in the Schumann–Runge[87] bands, resulting in dissociation of $O_2$ and production of atomic oxygen, O.
3. The $0.2-0.35 \, \mu m$ range. This is dominated by absorption by ozone, in two bands, again named after the scientists who first studied them.[88] The Hartley band has a wide peak from $200-310 \, nm$, while the Huggins band is a more structured region from 310 to 350 nm. There are also weak ozone absorption bands, the Chappuis bands, distributed between 450 and 850 nm, that is, mainly in the visible spectrum.

The most energetic photons are more likely to react when they encounter a molecule than are those with longer wavelengths,[89] and so are absorbed at the greatest heights. Figure 8.1 shows these three spectral regions and the height in the atmosphere at which the absorption is greatest within each of them.

A further common subdivision of the UV spectrum is that into the UV-C ($0.2-0.28 \, \mu m$), the UV-B ($0.28-0.32 \, \mu m$), and the UV-A regions ($0.32-0.4 \, \mu m$). This time the reason for the categorization are mainly biological; UV-C is the most, and UV-A the least, harmful to plants and animals. Virtually all of the solar UV-C is absorbed, mainly by oxygen, and most of the UV-B, by ozone, so a small fraction (exactly how much depending on the ozone amount present, the solar zenith angle, and other factors such as cloud cover) penetrates to the Earth's surface, along with all of the relatively benign UV-A.

[84] Ironically, tropospheric ozone can also be a health hazard itself when it reaches high concentrations, as it does in some heavily polluted urban environments.

[85] The bond in an $O_2$ molecule requires energy $E = 8.28 \times 10^{-19}$ J to break, corresponding to a photon of wavelength $hc/E = 0.24 \, \mu m$ or less. The bonds in $O_3$ are considerable weaker, and in theory can be dissociated by visible light.

[86] The term ionosphere is very common in the literature but can be confusing since, unlike the other atmospheric 'spheres', it does not refer to a well-defined vertical region, but to the layers (most of which are inside the upper mesosphere and the lower thermosphere) where high ion and electron densities are found.

[87] Victor Schumann (1841–1915) was an experimental spectroscopist who made observations of the UV spectrum of atmospheric gases with an instrument of his own design. Carle Runge (1856–1927) was a professor at the University of Hanover, where he worked on the theory of spectral line positions.

[88] The absorption by ozone of wavelengths shorter than 300 nm was discovered in 1880 by Sir Walter Noel Hartley of the Royal College of Science in Dublin, who went on to suggest that absorption by atmospheric ozone was responsible for limiting solar UV radiation at the Earth's surface. In 1882 James Chappuis in Paris discovered ozone absorption in the visible spectral region, originally in experiments to liquefy the gas, which was found to have a deep violet colour. In 1890 Sir William Huggins discovered a group of long-wave UV lines in the spectrum of the star Sirius, which was interpreted by Fowler and Strutt as absorption by terrestrial ozone.

[89] Photons with energies above the threshold required to dissociate or ionize common molecules experience continuum absorption, that is, absorption with a high probability at all wavelengths, since, unlike molecular vibration-rotation transitions (see Section 6.7) these processes are not quantized.

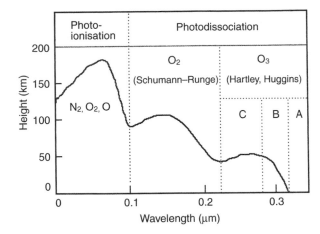

**Fig. 8.1**

The solid line is the height in the atmosphere at which UV radiation is most strongly absorbed, as a function of wavelength. Photons with wavelengths shorter than about 0.1 μm cause ionization of molecules and their dissociation products to occur, and produce free electrons, but do not penetrate much below the mesopause. Oxygen dissociation occurs in the Schumann–Runge band between 0.1 and 0.2 μm, and ozone dissociation occurs in the Hartley and Huggins bands between 0.2 and 0.3 μm. Note also the subdivision of the region in which ozone absorbs into UV-A, UV-B, and UV-C.

## 8.3  Photochemistry of ozone production

A simple intuitive scheme for the production of ozone is shown in Fig. 8.2. Ozone production is greatest in a layer centred on a height of about 25 km above the surface, simply because this is the where most of the photons capable of dissociating oxygen are absorbed. At higher levels, the pressure is low and there are too few oxygen molecules to intercept all of the photons. At lower levels, there are few photons left.

The details of the ozone production process were first set out by Sydney Chapman[90] in 1930. He considered a set of six reactions, three representing the production shown in Fig. 8.2:

$$O_2 + h\nu_1 \rightarrow O + O, \qquad [k_1(s^{-1})]$$
$$O + O \rightarrow O_2 \qquad [\text{slow}]$$
$$O + O_2 + M \rightarrow O_3 + M \qquad [k_2(\text{cm}^6 \text{ molecule}^{-2}\, s^{-1})]$$

and three representing possible paths for the destruction of ozone:

$$O_3 + h\nu_2 \rightarrow O + O_2 \qquad [k_3(s^{-1})]$$
$$2O_3 \rightarrow 3O_2 \qquad [\text{slow}]$$
$$O + O_3 \rightarrow 2O_2 \qquad [k_4(\text{cm}^3 \text{ molecules}^{-1}\, s^{-1})],$$

where $h\nu_1$ represents a photon in the spectral range that dissociates oxygen (about 0.1–0.2 μm, see Fig. 8.1), and $h\nu_2$ one in the range that dissociates ozone (about 0.2–0.3 μm). The rate constants $k$ describe how rapidly the reaction occurs; 'slow' means so slow as to be negligible compared with competing reactions.

[90] Sydney Chapman (1888–1970) in: A theory of upper-atmospheric ozone, *Memoirs of the Royal Meteorological Society*, 3(26), 103–25, 1930.

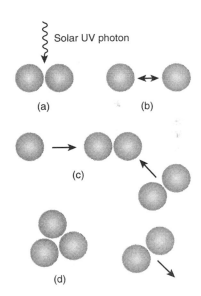

**Fig. 8.2**

A schematic of ozone production in the stratosphere, where each ball represents an oxygen atom. Part (a) shows an oxygen molecule; in (b) a solar UV photon of the right energy, corresponding to a wavelength between about 0.1 and 0.2 μm, dissociates the oxygen molecule to form two free oxygen atoms; in (c) one of the O atoms participates in a three-body collision with two $O_2$ molecules simultaneously; (d) the O atom combines with one of the $O_2$ molecules to produce $O_3$, while the other $O_2$ conserves energy and momentum.

The photons with wavenumbers $v_1$ and $v_2$ that fall inside the absorption bands of the $O_2$ and $O_3$ molecules respectively, have frequencies that correspond to allowed vibrations, and so tend to be absorbed by the molecules as discussed above. The difference here is that, as a result of the energy gained, the amplitudes of the vibrations are so large that not merely excitation, but actual dissociation of the molecule can result. The population of atomic oxygen that results from this dissociation can react with the remaining molecular oxygen,[91] producing ozone.

The ozone profile in the atmosphere represents the resultant of all of these reactions (assuming for the moment that there is no dynamical transport, and the Sun is permanently overhead). The different reactions occur at different rates, and the fastest reactions tend to dominate. Neglecting those labelled 'slow' and using the remaining four reactions, the rates of formation of O and $O_3$ can be expressed as:

$$\frac{d[O]}{dt} = 2k_1[O_2] - k_2[O][O_2][M] + k_3[O_3] - k_4[O][O_3]$$

$$\frac{d[O_3]}{dt} = k_2[O][O_2][M] - k_3[O_3] - k_4[O][O_3],$$

where the square brackets denote the abundance of the molecule. In the steady state

$$\frac{d[O]}{dt} = \frac{d[O_3]}{dt} = 0$$

so that

$$[O] = \frac{(2k_1[O_2] + k_3[O_3])}{(k_2[O_2][M] + k_4[O_3])}$$

and

$$[O_3] = \frac{k_2[O][O_2][M]}{(k_3 + k_4[O])}.$$

The rate constants for these reactions are temperature dependent, but typical values in the stratosphere are $k_1 \sim 10^{-12}$ s$^{-1}$, $k_2 \sim 2 \times 10^{-33}$ cm$^6$ molecule$^{-2}$ s$^{-1}$, $k_3 \sim 10^{-3}$ s$^{-1}$, and $k_4 \sim 10^{-15}$ cm$^3$ molecule$^{-1}$ s$^{-1}$. Using mean values for the concentrations of $[O] \sim 10^7$ molecule cm$^{-3}$ and $[O_3] \sim 10^{13}$ molecule cm$^{-3}$, and the values of the rate constants above we find that

$$k_3 \gg k_4[O];$$

hence

$$[O_3] \approx \frac{k_2[O][O_2][M]}{k_3}.$$

This result shows that the concentration of $O_3$ in the atmosphere depends primarily on the rate of ozone production in the three-body reaction between $O_2, O$, and M, and inversely on $k_3$, the rate constant for the photochemical destruction of ozone by solar UV.

[91] Because it involves two molecules combining to form a single species, this reaction can proceed only during a three-body collision, because otherwise energy and momentum cannot be conserved. The molecule $M$ providing this service can be any other atmospheric molecule, so statistically is most likely to be $N_2$ or $O_2$.

## 8.4 Variation of ozone concentration with altitude

To calculate vertical profiles of ozone abundance using Chapman's theory requires some additional work, (a) to incorporate the temperature dependence of the reaction rates $k$, which is considerable, and (b) to obtain the photon flux penetrating to each level in the atmosphere from a radiative transfer calculation like those discussed in Chapters 6 and 7.[92] A typical result appears in Fig. 8.3, and it shows that the Chapman scheme of reactions predicts ozone amounts that are far too large. The main reason is that, although the scheme gives about the right production rates for ozone, there are additional reactions, unknown in Chapman's time, on the removal side of the ozone budget.

## 8.5 Catalytic cycles

These involve reactions between ozone and several trace species in the stratosphere. These trace species, some of them of natural origin and some man-made (and some both), are present in such small amounts (a few parts per billion) that their importance was not realized until the 1970s. The crucial discovery was that the trace species act as catalysts, meaning that they accelerate ozone destruction reactions but are not themselves destroyed or consumed. This is why large amounts of ozone can be destroyed by tiny amounts of trace species, the so-called 'ozone depleting substances' or ODS.

The earliest of these to be identified were the oxides of nitrogen, NO and $NO_2$, collectively known as $NO_x$, which are produced naturally in the atmosphere by lightning and cosmic rays, and artificially in various ways including aircraft emissions. Even in Chapman's time, these were present in amounts large enough to shift the balance towards smaller amounts of ozone than his simple theory predicted. In recent decades, their amounts have been increasing due to human activities, including, it is thought, the increased use of artificial fertilizers designed to nitrify the soil; these release $N_2O$ into the atmosphere.

In the stratosphere, $N_2O$ reacts with atomic oxygen to produce NO, nitric oxide. The NO then combines with ozone to produce a molecule of ordinary diatomic oxygen and one of nitrogen dioxide

$$NO + O_3 \rightarrow NO_2 + O_2$$

The nitrogen dioxide reacts with more atomic oxygen, producing NO again

$$NO_2 + O \rightarrow NO + O_2$$

so the net reaction is the sum of these two:

$$O_3 + O \rightarrow 2O_2.$$

The direct combination of O with $O_3$ is less efficient than this chain, in which NO is destroyed and then restored, to start the cycle again with a new $O_3$ molecule. In this kind of catalytic behaviour, one NO molecule can be involved in the destruction of thousands of ozone molecules.

[92] Students wishing to try out the calculations for themselves could try visiting the Columbia University website, starting at www.columbia.edu/itc/chemistry/chem-c2407/hw/ozone_kinetics.pdf, where access is possible to a program that calculates the ozone profile using Mathematica and MS Excel.

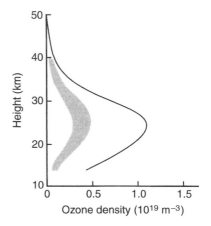

**Fig. 8.3**
Ozone abundance versus height (single line) calculated using Chapman's theory, compared to the range of observed profiles (shaded area). The discrepancy between theory and observation is nearly an order of magnitude.

While active nitrogen (so called to distinguish it from the inert form, $N_2$, that makes up most of the atmosphere) was the first ozone destroying species to be studied, and it remains an important one, we now know that the problem is more complicated. Several other species also catalyse the destruction of ozone, and some of them are more effective, and increasing in abundance faster, than $NO_x$. The greatest threat comes from certain chlorine, fluorine, and bromine compounds, especially the families known as chlorofluorocarbons (CFCs), hydrochlorofluorocarbons (HCFCs), and bromofluorocarbons (halons). These are very stable molecules that are used as refrigerants, solvents, foam blowing agents, fire extinguishers, dry cleaning, and a host of other applications. They were thought to be harmless, but in fact they, like oxygen, can be dissociated in the upper atmosphere by energetic UV photons. The resulting active halogen compounds are not stable at all, but virulently reactive, and with a great affinity for catalysing ozone destruction.

A common CFC, $CF_2Cl_2$ (popular name Freon 12) is dissociated in the stratosphere by UV photons in the reaction

$$CF_2Cl_2 + h\nu_3 \rightarrow CF_2Cl + Cl \qquad [k_5 = 1.0 \times 10^{-7}\ s^{-1}]$$

producing atomic chlorine which then reacts with ozone:

$$Cl + O_3 \rightarrow ClO + O_2 \qquad [k_6 = 2.1 \times 10^{-11}\ cm^3\ molecule^{-1}\ s{-1}]$$

and is regenerated

$$ClO + O \rightarrow Cl + O_2 \qquad [k_7 = 3.8 \times 10^{-11}\ cm^3\ molecule^{-1}\ s^{-1}]$$

To find how this affects the ozone abundance we have to add these reactions to the Chapman scheme and solve for $d[O]/dt, d[O_3]/dt, d[CF_2Cl_2]/dt, d[Cl]/dt, d[ClO]/dt$, thus:

$$\frac{d[O]}{dt} = 2k_1[O_2] - k_2[O][O_2][M] + k_3[O_3]$$
$$- k_4[O][O_3] - k_7[ClO][O]$$

$$\frac{d[O_3]}{dt} = k_2[O][O_2][M] - k_3[O_3] - k_4[O][O_3] - k_6[Cl][O_3]$$

$$\frac{d[CF_2Cl_2]}{dt} = -k_5[CF_2Cl_2]$$

$$\frac{d[Cl]}{dt} = -k_5[CF_2Cl_2] - k_6[Cl][O_3] + k_7[ClO][O]$$

$$\frac{d[ClO]}{dt} = k_6[Cl][O_3] - k_7[ClO][O]$$

whence, in the steady state,

$$[O_3] = \frac{k_2[O][O_2][M]}{(k_3 + k_4[O] + k_6[Cl])}.$$

Again, knowing the reaction rates and calculating the profiles of O and Cl from the relevant photon fluxes as a function of height, the expected ozone profile can be calculated. The amount at each level is less than in the Chapman scheme because of the extra term in the denominator; we also note that the reaction rate $k_6$ for Cl is about 10,000 times bigger than $k_4$ for atomic oxygen, so small amounts of chlorine have a large effect. When the nitrogen, bromine, and other cycles are all added in, the calculated ozone profile is close to the mean observed profile, and the discrepancy noted in Fig. 8.3 can be accounted for.

## 8.6  Ozone measurements

Figure 8.4 shows a plot of the total amount of ozone in a column of atmosphere as a function of latitude and season, averaged over a 5-yr period. These data were obtained by the Total Ozone Mapping Spectrometer (TOMS), a satellite instrument that works by measuring the amount of solar UV radiation that is scattered back into space from the stratospheric ozone layer. This technique gives only poor vertical resolution, but is very good for measuring the total ozone in a column under the spacecraft. Infrared instruments like those described in Chapter 9 can also measure ozone profiles, but, because the profiles contain a lot of important detail that remote sensing cannot resolve, direct *'in situ'* measurements are still particularly important for understanding the distribution and depletion of ozone.

Meteorological balloons or 'radiosondes' are used routinely to obtain data, principally on pressure, temperature, and humidity, for weather forecasting. Sites at about 50 locations around the world also make regular ozone vertical profile measurements using the *ozonesonde*, a small instrument that is mated to a conventional radiosonde. The ozonesonde sensor uses an electrochemical cell containing platinum electrodes immersed in potassium iodide solution. This reacts with ozone when ambient air is pumped through the device as the balloon ascends, allowing a current proportional to the ozone concentration of the sampled air to flow. The measured value of the current is telemetered to the ground, along with values for the air flow

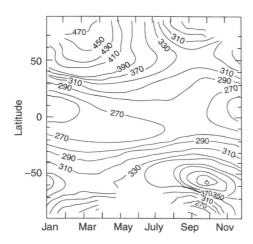

**Fig. 8.4**

Total column ozone amounts as a function of latitude and time of year, showing the mean seasonal fluctuations from 1979 to 1984 as measured by a NASA satellite instrument called the Total Ozone Mapping Spectrometer. The units are 0.01 mm of ozone at STP ('Dobson' units).

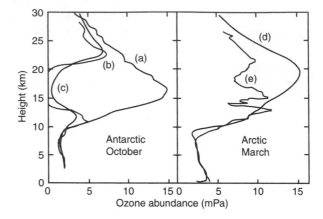

**Fig. 8.5**

Ozone profiles in Spring in Antarctic (left) and Arctic (right) obtained with ozonesondes. The Antarctic profiles show the partial pressure of ozone as a function of height (a) before (averages for October 1962–1971) and (c) after (averages for October 1992–2001) the ozone hole began its annual appearance, plus an individual profile (b) obtained on 2 October 2001, when ozone depletion was at its maximum. The Arctic average for October 1988–1997 (d) shows little evidence for depletion, but 30 March 1996 (e) shows that less severe 'bad ozone days' occur in the northern hemisphere also. (Data from NOAA.)

rate, air pressure, and pump temperature, and the ozone concentration in the air is determined from a standard formula provided by the manufacturer. Ozonesondes typically ascend to altitudes of about 35 km (about 4 hPa) at mid latitudes and somewhat lower at the south pole, where they are used to check the progression of the annual ozone hole (Fig. 8.5).

## 8.7  The Antarctic ozone hole

The Antarctic land mass, completely surrounded by ocean, has a symmetry that is reflected in the formation of a large, stable polar vortex that isolates the air inside it. As a result, stratospheric temperatures there are the coldest anywhere on the planet, falling to 170K or even less. Similar conditions do not exist over the north pole, where the vortex is less stable and more mixing with warm air from outside the Arctic Circle occurs, so temperatures are generally higher. Thus wintertime temperatures in the Arctic stratosphere are not persistently low for as many weeks as over Antarctica, which results in correspondingly less ozone depletion in the Arctic.

[93] The principal component is nitric acid trihydride, $HNO_3 \cdot 3H_2O$, known as NAT.

When the Antarctic temperatures fall below 185K, the formation of *polar stratospheric clouds* (PSCs), composed primarily of frozen nitric acid crystals,[93] occurs. This has two crucial effects: first to remove most of the active nitrogen that is normally present in the air, and second to provide solid surfaces to catalyze some complex heterogeneous chemistry. The chemically active chlorine and bromine that is normally trapped as relatively stable nitrates then can attack ozone in photochemical reactions that occur rapidly when sunlight returns to Antarctica in September and October of each year. The resulting ozone 'hole' has grown both wider and deeper in recent decades (Fig. 8.6), to the extent that virtually no ozone is present at some heights, and more than 60% of the total ozone inside the

**Fig. 8.6**
Schematic showing the extent of the ozone 'hole' over the Antarctic on 11 September 2003. The area covered on this day was one of the largest on record—about 25 million km$^2$, more than the area of Antarctica and about three times that of Australia.

midwinter polar vortex is lost, although of course it recovers later in the year. This deepening cycle reflects the growth in stratospheric pollution, with a built-in delay of a decade or more due to the slow transfer of chlorine and other compounds through the tropopause. Recent reductions in CFC release, due to new laws on their use, have reduced the rate of ozone loss but are subject to the same delay.

## 8.8 The global distribution of ozone

The vertical profile and the total amount of ozone present at any particular location on the globe depends on the balance between production and destruction, which in turn depend on three main factors: (i) the intensity of the solar flux; (ii) mean temperature; (iii) transport, especially in the meridional (N–S) direction by the high winds that prevail in the stratosphere; and (iv) the distribution of pollutants involved in the catalytic destruction of ozone.

Ozone production is greatest in the tropics where more solar photons are available. However, the prevailing winds in the stratosphere tend to transport the ozone polewards to latitudes where the rate of natural ozone destruction is less. As a result, ozone is generally found in higher concentrations at high latitudes (except of course inside the Antarctic ozone hole where special conditions apply—see below). Because of the temperature dependence of the reactions involved in the production and destruction of ozone, the abundance also tends to be larger in the winter than the summer.

Most of the man-made halogen-containing gases involved in ozone destruction are released in the more densely populated northern hemisphere: about 90% in the latitudes containing the industrialized countries of Europe, Russia, Japan, and North America. Since they are chemically stable and almost insoluble in water, they are mixed fairly rapidly around the globe in the lower atmosphere by winds and turbulence. The vertical transfer into the stratosphere takes longer, many years, and occurs mainly in tropical latitudes, but once in the stratosphere global mixing again occurs quite quickly

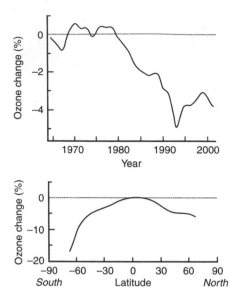

**Fig. 8.7**
Global total ozone has been in decline
since about 1980 (top); the particularly
sharp dip in the early 1990s followed the
eruption of Mt Pinatubo, which produced a
large, temporary increase in the aerosol
content of the stratosphere. In the bottom
panel, ozone changes between 1980 and
2000 are compared for different latitudes,
showing that most of the change had been
in or near the polar regions, especially the
Antarctic. (Data from National
Oceanographic and Atmospheric
Administration.)

and all latitudes contain nearly equal amounts of chlorine and bromine.
However, ozone depletion is greatest in Antarctica because of the unique
meteorological conditions there that lead to PSC formation.

Averaged over the globe, stratospheric ozone depletion since 1980
amounts to about 4% of the total ozone amount present worldwide
(Fig. 8.7, top). If this loss is plotted as a function of latitude, we see that
most of it is occurring near the poles, especially the regime around the
Antarctic affected by the ozone hole (Fig. 8.7, bottom). Mid-latitude depletion is also due to the increase in atmospheric concentrations of human-made
compounds, but the reactions proceed more slowly without the powerful
catalytic effects of PSC formation.

The stratosphere over the Arctic region in the north is typically some
10K warmer than its counterpart in the south, and PSC formation is less
common and tends to be short-lived. Transient effects are seen at all latitudes following large volcanic eruptions, when large amounts of volcanic
particles injected into the stratosphere mimic some of the effects of PSCs
in accelerating the effects of pollution on ozone, before settling out after
a few years. The effect of the eruption of Mt Pinatubo in 1991 is clearly
visible in Fig. 8.7.

## Further reading

*Atmospheric Chemistry and Physics*, J.H. Seinfeld and S.N. Pandis, Wiley-
Interscience, New York, 1998.
*Introduction to Atmospheric Chemistry*, Daniel J. Jacob, Princeton University Press,
Princeton, 1999.
*Atmospheric Change: An Earth System Perspective*. T. Graedel and P. Crutzen,
W.H. Freeman & Co., New York, 1995.
*Executive Summary of Scientific Assessment of Ozone Depletion: 2002* (PDF,
476K). Available on the web site of the National Oceanic and Atmospheric
Administration (NOAA) www.unep.org/ozone/pdf/05-ExecutiveSummary.pdf

# Questions

1. Outline the classical 'Chapman' theory for the formation of ozone in the Earth's stratosphere, and write down the equations for the main reactions involved.
2. Show that, with suitable simplifications, the scheme can be used to derive an approximate expression for the number density of ozone at a given height in the stratosphere as a function of the number density of atomic and molecular oxygen and other relevant quantities.
3. Explain the importance of the photochemical rate constants involved, which solar wavelengths are important, and why chemically inert gases, especially $N_2$, have a role in the scheme.
4. Indicate in general terms what discrepancies are found when the ozone profile calculated by Chapman's theory is compared to measurement. What are the main causes of these discrepancies?
5. What are the principal reactions that need to be added to Chapman theory to achieve a more realistic result?
6. What are the principal mechanisms involved in the production of the Antarctic ozone 'hole'? Is there an Arctic ozone hole? If not, why?

# Climate observations by remote sensing

## 9.1 Introduction

Progress with understanding any aspect of the Earth as a physical system, in particular the environmental effects of climate change, depends critically on our ability to design and carry out suitable programmes of *measurements*. Theory and modelling alone cannot improve our understanding of the physics on which predictions rely, and only relevant data of high quality can test and improve specific models and parameterizations. By developing suitable instruments, we can explore and understand the physical processes in the atmosphere and oceans at the same time as we monitor the climate system for variations and secular change.

Most of the existing body of knowledge about the climate system, and the evidence for global change, comes from traditional instrumentation deployed on ships, aircraft, and balloons, plus that housed in meteorological stations of various sizes, ranging from small automatic stations to large manned bases at locations ranging from city airports to the polar regions. These provide a long baseline in time, but face the problem that truly global quantities are very difficult to obtain from discrete data, especially when the sampling is patchy. The relatively recent technique in which spectroscopic measurements of the radiation emitted by, or reflected from, the Earth are made by Earth-orbiting satellites and interpreted in terms of temperature, energy budget, composition, and cloud field, has revolutionized climate research.

*Remote sensing* is the popular name for the measurement of physical quantities at a distance by spectroscopic methods. Its application to the study of the Earth's atmosphere and surface from satellites is a rapidly growing field, with many applications, and new or improved techniques are constantly being developed. In this chapter we look at some of the basic principles and technical challenges that are involved, and at the use of remote sensing to understand the processes at work in the climate system, and to detect global change.

## 9.2 Ground-based measurements

Satellite-based measurements are important because they offer the global coverage that is often so essential, but they do not replace ground-based measurements completely. Instruments on the ground can be continuously maintained and checked against external calibration standards, and so they give the best performance for long-term monitoring of subtle changes in climate variables.

It might seem at first to be simple to obtain a mean global surface temperature by taking readings from meteorological stations scattered around the world, and combining these into a spatial and seasonal average. However this is not trivial, since the distribution of measurement is non-uniform in space and time and confusion between seasonal and non-secular changes can easily occur. Also, when long runs of data are used to search for trends, problems of changes in measurement techniques and unreliable intercalibration are encountered. For example, measurements made within the urban 'heat island' are generally higher than those made at airports or in other more open situations, and the last 50 yrs has seen the gradual move of met stations from one to the other.[94]

Remote sensing, as well as traditional methods using thermometers, barometers, and rain gauges, may be used to collect data at ground based stations. Since the pioneering work in the 1920s by Dobson,[95] who studied stratospheric ozone using UV spectroscopy, it has become possible to study an increasing number of stratospheric variables up to 50 km above the surface, with suites of instruments that remain fixed on the ground. This kind of vertical coverage is essential since the stratosphere contains the main source for some greenhouse gases (e.g. ozone), and the main sink for others (e.g. methane). Predicting the rate of global change depends on understanding processes there, and furthermore most climate models predict that such change can be seen first in its effects on the stratosphere. In particular, cooling of the stratosphere in response to an increase in the $CO_2$ mixing ratio is greater than the corresponding heating at the surface and therefore easier to detect. Global cooling of the stratosphere has been seen in satellite data since the 1970s.

Global coverage from the surface requires international networks of ground stations. A good example is the Network for the Detection of Stratospheric Change, which commenced operation in January 1991. The NDSC consists of about 20 primary stations and a larger number of complementary stations covering the globe (see www.ndsc.ncep.noaa.gov). These were designed to address the rate of build-up of greenhouse gases in the atmosphere as a whole by making upper-air measurements, since those at the surface are confused by local effects such as the concentration of pollutants near population centres, rainout and so forth. The measurements made by the NDSC vary from site to site, for budgetary more than scientific reasons, but the following are brief descriptions of some of the most important.

The *differential absorption lidar* technique uses two or more adjacent wavelengths (0.353 and 0.308 μm for ozone, for example), one of which is much more strongly absorbed by the species of interest than the other. The returned signals are small but, given cloud-free skies and a long integration time (up to several hours) night-time profiles from 5 to 50 km with vertical resolution of 200 m and absolute errors of less than 10%, can be achieved.

*Temperature measurements* by lidar rely on Rayleigh scattering from air molecules, typically at the wavelength of Nd lasers (0.532 μm). The signal strength is proportional to the air density and hence can be related to the temperature. Profiles are obtained from about 30–80 km altitude, yielding

[94] Tales abound of data that, on enquiry, turned out to have been made by rain gauges that were installed upside down, or temperature measurements that were made indoors; while hopefully most of these tales are apocryphal they still make the point that measurements not made in exactly the same way at every location and time are potentially misleading.

[95] Gordon M.B. Dobson (1889–1976) was Reader in Meteorology in the Clarendon Laboratory, Oxford, from 1920 until 1960. He pioneered atmospheric remote sensing in the 1920s with the Dobson ozone spectrometer, which is still in use worldwide to this day.

temperature to an accuracy of better than 2K with 200 m vertical resolution. Below 30 km, backscatter from atmospheric aerosols confuses the temperature signal and below about 26 km this contribution is so dominant that profiles of the aerosols themselves can be derived. Lidar instruments observing the backscatter from water vapour at 408 nm obtain vertical profiles from near the surface to the lower stratosphere.

*Infrared spectroscopy* of the key greenhouse gases, their source and sink species and their reaction partners using their vibration-rotation bands in the 2–15 μm wavelength range, is possible simultaneously with a Fourier spectrometer such as a Michelson interferometer. These achieve spectral resolutions in the region of 0.002 cm$^{-1}$, which is sufficient to separate the stratospheric from the (often much larger) tropospheric contribution to the spectral lines of interest. The Sun is observed over a range of solar zenith angles to obtain a range of air masses to observe weakly and more strongly absorbing lines with the same set-up, improving the vertical resolution.

For certain species, including ozone, nitrogen dioxide, nitrogen trioxide, and chlorine and bromine oxides, *ultraviolet (UV) and visible spectroscopy*, basically an update of Dobson's technique for stratospheric ozone monitoring, gives superior results to the infrared. Grating instruments with a spectral resolution on the order of 1 nm and integration times of 20 s to 20 min are used, and precisions of around 1%, for $O_3$ and $NO_2$, and around 10%, for the other species, are claimed.

Thermal emission from optically thin stratospheric thermal emission lines are monitored using *microwave* receivers operating at wavelengths around 22.2 GHz (for water vapour), 110 GHz (for ozone), and 279 GHz (for chlorine oxide). High spectral resolution defines the shapes of the pressure-broadened lines and allows vertical profiles to be retrieved, but with less vertical resolution than is obtainable with active techniques such as lidar. Profiles with a resolution of around one scale height (6–8 km) are obtained for 30 min integration times for water vapour (20–85 km), ozone (20–75 km), and chlorine oxide (25–45 km).

## 9.3  Satellite measurements

Instruments on satellites in high-inclination[96] orbits can measure the net reflected and thermally emitted radiation as a function of latitude, longitude, time of day, and season, covering the whole planet (Fig. 9.1). They

[96] The *inclination* of the orbit is the angle between the plane of the orbit and the equatorial plane of the planet. A 'high-inclination' orbit is one that is close to 90°, which means that the satellite crosses the tropics at an angle perpendicular to the equator, and passes over both poles.

**Fig. 9.1**
Climate observations from space involve remote sensing using spectrometers and radiometers mounted on a satellite orbiting typically 500–1,000 km overhead. Reflected sunlight (left) or thermal emission (right) may be the source of the radiation observed, depending on the property to be measured.

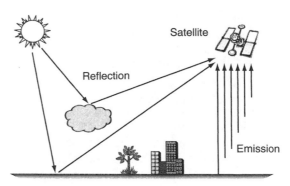

can measure vertical temperature profiles, minor constituent distributions (e.g. water, ozone), cloud cover, thickness, and height, and ocean temperature. Individual measurements or profiles typically take a second or less to acquire, and can be repeated as the scene changes below, allowing the parameters to be mapped in three dimensions. Maps from successive orbits, approximately every hour, makes it possible to study processes at work in large-scale weather systems and photochemical regimes like the stratospheric ozone layer. Because the same instrument is used everywhere, problems of intercalibration are largely eliminated.

The definition of remote sensing includes imaging from space, using high-resolution photography and television, often limited to the visible part of the spectrum. However, the more interesting case where quantitative spectroscopy or radiometry is interpreted in terms of vertical *profiles* of key climate variables, such as atmospheric temperature or ozone abundance, the technique (often called *remote sounding* in this case) involves advanced optical methods and sophisticated mathematical approaches to information retrieval, some of which we will discuss below.

The most basic remote sensing observations are those of the energy budget of the Earth, from an investigation of the balance between the ingoing and outgoing fluxes of radiative energy that drive the thermodynamic engine that regulates the climate. Going beyond that, infrared spectroscopy and radiometry from space platforms are now routinely used to measure the temperature and composition of atmospheric gases and solid surfaces that constitute the atmosphere, oceans, and land surface. The stratospheric ozone layer, pollution in the lower layers of the atmosphere, and the 'greenhouse' gases that drive global warming, are all conveniently studied in this way. Mineralogy, vegetation (including agricultural products and some of the species in the sea), the icy cryosphere, and volcanism and its products are also accessible to sensors in space. Radar and lidar are rapidly being adopted to investigate land and ice topography, sea-state, and to infer wind fields.

Operationally, remote sensing is used to observe the weather, snow cover, crop development, and a range of natural resources. In some cases, current remote sensing instruments and methods are too crude to replace older ways of obtaining data completely. Satellite measurements of atmospheric temperature and humidity near the surface are not good enough to supersede the traditional balloon-borne radiosonde for providing the input to weather-forecasting models, except of course where the latter is not available (which is, however, a good deal of the time over most of the globe). Current meteorological satellite sensors lack vertical resolution and can be confused by clouds, but their use has radically improved weather forecasting and further progress is rapid.

The physics of remote sensing was developed in earlier chapters. The Earth, suspended in space, emits a varying stream of electromagnetic radiation in all directions at wavelengths ranging from the UV to the microwave. On the night side of the planet, the photons that make up this energy flux originate as thermal emission from the Earth's surface and atmosphere, and their intensity and wavelength distributions are functions of the composition and physical state of the emitter. On the daylit side of the globe, there are additional photons of solar origin that have been reflected from the surface or from clouds, and scattered from aerosols or, at the shorter wavelengths, from the molecules of gas that make up the atmosphere. Overall about one-third of the flux from the Sun, on average, is reflected directly back to space. The rest is absorbed, its energy reappearing later as long-wave infrared and microwave emission.

Although the Earth, on average, radiates about the same total energy to space as a 255K blackbody, the terrestrial spectrum is rich in molecular vibration-rotation lines (Section 7.2.1), which contain information about the physical state of the atmosphere, and continuum fluctuations that represent weak and/or complex atmospheric bands plus surface and cloud emissivity variations. The most pronounced of these spectral features are those caused by absorption and emission in the bands of the principal atmospheric minor constituents, especially water vapour and carbon dioxide. The backscattered and reflected photons also carry information about the planet as a physical system, and about the human environment below. In each case, the information we seek is coded into the intensity as a function of wavelength, and can be acquired by measuring the relevant parts of the electromagnetic spectrum of the Earth with instruments that are sensitive, but also have to be compact, robust and reliable, since they are to be launched into space.

## 9.4   Infrared instruments for remote sounding

[97] Radiometric (radiation-measuring) instruments are generally of two types: radiometers, that generally have low spectral resolution but high absolute radiometric accuracy; and spectrometers, that resolve spectral bands (and sometimes individual spectral lines) without necessarily being radiometrically calibrated. Devices that have all of these properties combined are called spectroradiometers.

A typical satellite radiometric instrument[97] (real examples are discussed later) will have the components and general layout shown in Fig. 9.2. These gather, focus, modulate, filter, detect, amplify, and calibrate the signal from a well-defined field of view in the atmosphere, as we describe below, taking each subsystem in turn.

### 9.4.1   Calibration

The radiometer views the Earth through a plane scan mirror, which can be rotated to view two calibration targets. One of these is a view sideways to cold space (zero radiance), the other is an on-board blackbody at a known temperature $T_{BB}(T_{BB} > T_{Earth} > T_{space})$. By design, most instruments have a linear response, that is, the signal at the output is proportional to the radiance $R$ entering the aperture and a plot of signal $S$ against incoming

**Fig. 9.2**
A schematic diagram of an infrared radiometer, showing the main subsystems, which are discussed individually in the text.

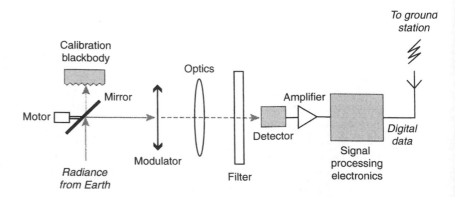

radiance $R$ is a straight line. If we define an overall gain factor $g$ (discussed further in Section 9.5.1 below) and an offset $S_0$, $S$ is given by:

$$S = gR + S_0$$

when viewing the Earth, and

$$S = gB(T_{BB}) + S_0$$

when viewing a calibration blackbody at a known temperature $T_{BB}$, and just

$$S = S_0$$

when viewing cold space. Thus, regular rotation of the scan mirror to view cold space and the blackbody calibration target allows the two constants $g$ and $S_0$ to be determined, and checked frequently during measurement sequences and throughout the lifetime of the instrument.

The in-flight calibration system is itself calibrated in the laboratory before launch, to check the linearity of the radiometer and other factors such as the emissivity of the on-board blackbody.

The latter may in reality be somewhat less than unity, since simple plates or cones painted black are often used to save weight. Their properties have to be checked against a sophisticated laboratory blackbody, with near-perfect emissivity and a controlled, adjustable temperature. This provides a variable, well-defined input radiance, within 99.99%, of that from an ideal blackbody, as a standard (Fig. 9.3).

## 9.4.2 Modulation

Modulation is a common feature of systems designed to measure radiance accurately (as opposed to, for example, imaging systems, which are not required to deliver absolute values for the intensity). The modulator or 'chopper' is simply a rotating or vibrating blade that regularly interrupts

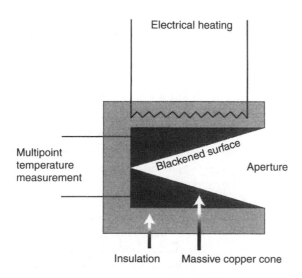

**Fig. 9.3**
Schematic of a practical calibration blackbody. Radiation entering a deep cone cannot exit without making many reflections; the inside can be coated with special black paint that has a reflectivity of less than 0.01. The use of massive amounts of copper and insulation ensures that the temperature is constant all over the cone.

the beam, chopping the beam into pulses of light that generate an alternating signal at the detector at a frequency typically 100–1,000 Hz. An a.c. signal is easier to amplify, and can be filtered electronically to remove the background signal from the warm parts of the inside of the instrument, which of course glow brightly in the infrared, and to reduce the noise bandwidth.

### 9.4.3  Radiometer optics

These focus the energy onto the detector, and define the field of view of the instrument. Often there is an intermediate focus for the chopper, to keep it small. In infrared instruments, reflecting optics are used where possible because transmitting components such as lenses are limited in the wavelength range they can cover. Some sophisticated (and expensive) materials (e.g. germanium and diamond) transmit over a wide range and are used where lenses cannot be avoided, such as the field lens in front of the detector.

The aperture must be large enough to avoid problems with diffraction, otherwise the field-of-view will include undesirable side lobes due to the diffraction pattern that results from the finite size of the aperture. The Rayleigh criterion states that, to just resolve a feature of angular width $\theta$ (the diameter of the first Airy disc) using an aperture of diameter $D$ at wavelength $\lambda$ we must have

$$D \geq \frac{1.22\lambda}{\theta}.$$

For limb sounding (Section 9.6), the region viewed is typically $\sim$2,500 km away, which means an angular resolution of $\sim$0.5 mr to resolve 1 km at the limb. In the thermal infrared, $1.22\lambda \sim 20$ μm so by this criterion the aperture must be $\gg$4 cm or say $\sim$40 cm to avoid diffraction effects. This is quite a large telescope to mount on a spacecraft, and contributes to a very large instrument; the High-Resolution Dynamics Limb Sounder (HIRDLS), an example of a state-of-the-art limb sounder (Section 9.7.3), is more than 1 m on a side and weighs nearly 200 kg.

### 9.4.4  Interference filters

These select the set of channels or wavelength ranges $\Delta\lambda$ to which the instrument is sensitive. The principle by which they work is basically that of the *Fabry-Perot etalon*,[98] which in its original form consisted of two partially reflecting surfaces which are precisely parallel and separated by a distance $d$, illuminated by a parallel beam of radiation, as shown in Fig. 9.4. Because of interference between the multiply reflected beams inside the etalon, its throughput as a function of wavelength is periodic, with a regular series of maxima and minima in transmission, as shown.

The theory of the etalon is covered by most basic books and courses in elementary optics. Recapping briefly, if $a_0$ is the amplitude of the incoming beam and $r$ and $\tau$ are the reflectance and transmittance of a plate, then the

[98] First described by C. Fabry and A. Perot in *Ann. Chim. Phys.*, **16**, 115, 1899 and used by Perot in 1913 to demonstrate the presence of strong absorption features of ozone in the solar spectrum.

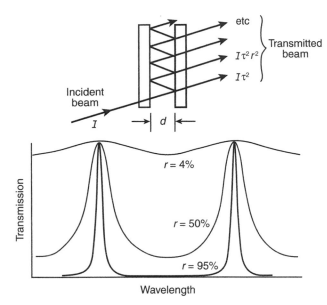

**Fig. 9.4**
(Top) Interference of multiple beams in a thin film with semi-reflecting interfaces (reflectivity $r$ and transmission $\tau$). (Bottom) The resulting net transmission as a function of wavelength of a single etalon for three values of $r$. In an interference filter, a large number of patterns similar to this, but with different periodicities, are superimposed such that all of them have a maximum that coincides at the wavelength it is desired to select.

transmitted amplitude is

$$a = a_0 \tau \{1 + r e^{-i\delta} + r^2 e^{-2i\delta} + \cdots \}$$

$$a = a_0 \tau \{1 - r e^{-i\delta}\}^{-1}.$$

The transmitted power is

$$\mathcal{T} = aa^* = \frac{a_0^2 \tau^2}{\{1 + r^2 - 2r \cos \delta\}},$$

which on manipulation gives *Airy's formula*:

$$\mathcal{T} = \frac{\mathcal{T}_0}{\{1 + F \sin^2 \left(\frac{\delta}{2}\right)\}},$$

where $\mathcal{T}_0 = a_0^2 \tau^2 / (1 - r^2)$ is the maximum transmission, and $F = 4r/(1 - r)^2$ is called the *finesse*.

The transmission $\mathcal{T}$ is periodic with maxima, called fringes, which occur when $\delta/2 = m\pi$, that is, when $nd \cos \theta = m\lambda$ for $m = 1, 2, 3, \ldots$ The integer $m$ is called the order of the fringe. The width of a peak as a fraction $\Delta m$ of $m$ is found by putting $\mathcal{T} = \mathcal{T}_0/2$ into Airy's formula; whence (for normal incidence)

$$F \sin^2 \left(\frac{\delta}{2}\right) = F \sin^2(\pi \, \Delta m) \approx F (\pi \, \Delta m)^2 = 1$$

and

$$\Delta m = \frac{1}{\pi \sqrt{F}}.$$

Thus, the sharpness of the maximum, or the spectral resolution if the etalon is being used for spectrometry, can be improved by increasing the finesse. This involves increasing $r$ at the expense of $\mathcal{T}$, and $\mathcal{T} \to 0$ to zero as $r \to 1$.

If several etalons are placed in series, some of the maxima and minima block each other and the combination can be made to have almost any desired transmission characteristic, including a single pass band.[99] In the interference filter, the interfaces between up to a hundred, or even more, thin layers of two or more kinds of infrared-transmitting dielectric material (such as cadmium or zinc telluride or selenide) are deposited on a substrate of infrared-transmitting material to make a stack of Fabry-Perot etalons that, in combination, define the desired pass band. Outside this band, filters emit radiation at their ambient temperature and so are often cooled to cut down the background flux of thermal radiation on the detector.

### 9.4.5   Thermal infrared detectors

Thermal detectors respond to the heating effect of the radiation using familiar physical effects. The three most common are:

(1) the variation of electrical resistance with temperature (devices using this effect are called *bolometers*),
(2) the temperature sensitivity of the polarization of the charge distribution in certain crystals (the *pyroelectric* effect), and
(3) the thermoelectric effect in which a temperature-sensitive potential appears across the junctions between dissimilar metals (thermo-couples). Large numbers of thermocouples can be connected in series to form a *thermopile*; modern thermopiles have thousands of bismuth–antimony thermocouple junctions in a very small mass and so can have a good responsivity with a small response time.

Thermal detectors are less sensitive than photon detectors (see below), but they cover a wider wavelength range and usually do not require cooling to operate at close to their maximum sensitivity.

### 9.4.6   Photon detectors

These detectors respond to individual photons by exciting a transition in a semiconductor. There are two distinct modes of operation:

(1) photoconductive devices, which measure the change in current due to carriers excited by the incident photons into the conduction band when the detector has a bias voltage across it, and
(2) photovoltaic, which measure the induced voltage in a detector without voltage bias.

Both types require cooling to suppress the thermal excitation of transitions, which would otherwise completely mask any contribution by the photons, particularly in the infrared where the energy carried by each photon is relatively small. Modern photon detectors are often fabricated as tiny elements only a few wavelengths across, in large arrays containing a million or more such elements, read out by a CCD just as in an ordinary digital camera.

[99] 'Blocking' materials, that absorb large segments of the unwanted part of the spectrum, may also be included to make it easier to define a single passband $\Delta\lambda$ in this way.

### 9.4.7  Electronics and telemetry

Apart from the use of low-noise preamplifiers to capture the very small signal from most types of detectors, the rest of the signal chain is fairly conventional. The signal is amplified, digitized, and transmitted back to the ground station, where the data from the instruments is relayed to the scientists who use it, along with housekeeping data (related to instrument health) and navigational data (so the observations can be related to position on the Earth).

## 9.5  Radiometric performance

The performance of remote sensing instruments is always being pushed to the limit. Detecting global warming, for example, requires temperature measurements that are accurate to better than 1K, and even routine meteorological data need to be of similar precision. The change in radiance for a 1K change in temperature from a source at 300K or less is not very large at any wavelength, and an instrument observing this emission from a distance of perhaps 1,000 km does not receive a very large signal. For the experimental climate physicist, therefore, it is crucial to be able to analyse the design of an observing system in terms of its *signal to noise ratio,* in order to understand what can and cannot be measured, and what the error budget is.

### 9.5.1  Signal

The signal $S$ at the output of the radiometer is usually designed to be a linear function of the radiance $R_\lambda$ at the input aperture. It is therefore a simple matter to calculate the relationship between the two.

The detector and its associated electronics give an output in volts for each watt of energy it receives. To get the figure in watts, first the radiance (in units of $W\,m^{-2}\,sr^{-1}\,\mu m^{-1}$) is multiplied by the filter transmission $a_\lambda$ at each wavelength and integrated over wavelength, then multiplied by the product of the input aperture of the telescope $A$ and its angular field of view $\Omega$, with units of $m^2$ and sr (steradians) respectively.[100] The conversion from watts to volts is achieved by multiplying by a linear electronic gain factor $g'$. Since normally the filter has a non-zero transmission over only a small range of $\lambda$ (see the examples in Fig. 9.8 and 9.13), the radiance can be assumed to be nearly constant with mean value $\overline{R}$. If we also replace the integral over the filter response, $\int_0^\infty a_\lambda\,d\lambda$ with the equivalent width of the filter $\Delta\lambda$ (i.e. treating it as a rectangular profile of width $\Delta\lambda$ with $a_\lambda = 1$ within this range and zero elsewhere) we have;

$$S = g'A\Omega \int_0^\infty a_\lambda R_\lambda\,d\lambda + \text{const}$$

$$\approx g'A\Omega\overline{R}\Delta\lambda + \text{const}$$

$$= g'\overline{R} + S_0,$$

[100] The product of aperture area $A$ and angular field of view $\Omega$ is called the *etendue* or *energy grasp* of the system. A useful rule of thumb for estimating the area $A_d$ of the detector, in an instrument performance problem where it is not given, is to use the well-known rule of optics that the etendue is conserved throughout the system. Then, if $A\Omega$ is known at the entrance aperture, $A_d \approx A\Omega$ for most systems, because the field of view of most detectors is made as large as possible, to keep the detector small. A solid angle of $\sim$1 steradian is about the largest practical value.

where $g'$ now incorporates all of the constants of the system and $S_0$ is a constant signal offset. As noted in Section 9.4.1 on calibration, these can be determined empirically when the radiometer views cold space (zero radiance, so $S = S_0$) and the on-board warm blackbody at temperature $T_{BB}$, so $S = gB(T_{BB}) + S_0$.

### 9.5.2 Noise

Detectors of all types produce random noise along with the signal, which limits the accuracy to which the signal can be measured. The noise is generated by certain fundamental processes, in particular the following three, one of which will normally dominate for any given experimental arrangement.

1. *Thermal or Johnson Noise* is caused by the thermal agitation of the charge carriers in any resistor, including the resistive component of many detector types. It exists even when there is no current and only disappears if the temperature is absolute zero, or in circuits where the resistance is vanishingly small. The root-mean-square noise voltage is given by

$$\bar{v}_{rms} = \sqrt{4kTR\Delta f},$$

where $k$ is Boltzmann's constant, $T$ is temperature (K), $R$ is the resistance ($\Omega$) of the circuit containing the detector, and $\Delta f$ is the frequency bandwidth (Hz), equal to the reciprocal of the length of time $\Delta t$ the signal is measured, called the dwell time (s). Note that the amplitude of Johnson noise depends only on the frequency bandwidth and not on the frequency of the signal itself; this is called 'white' noise.

   As is apparent from this expression, there are three ways to reduce thermal noise. The first is to narrow the bandwidth $\Delta f$, by integrating the signal for longer; the second is to reduce the number of resistive elements in the detector circuit, and the third is to reduce the temperature of the electronic components including the detector element itself.

2. *Shot Noise* occurs wherever electrons cross a junction, due to the fact that the current is made of finite discrete particles rather than a continuous flow of charge. The rms current fluctuation is given by

$$i_{rms} = \sqrt{2ie\Delta f} = \sqrt{\frac{2ie}{\Delta t}},$$

where $i$ is the current (A), $e$ is the change on the electron, and $\Delta f$ is again the bandwidth (Hz). Shot noise is also independent of the signal frequency.

3. *Photon Noise* is caused by the statistics of the photon flux, so it is similar to shot noise and given by the same expression with current replaced by photon flux $\Phi$:

$$i_{rms} = \sqrt{2\Phi e\Delta f} = \sqrt{\frac{2\Phi e}{\Delta t}}.$$

Photon noise can be reduced by limiting the background photon flux with cold shields around the detector, or by cooling the whole instrument. A further reduction can be obtained by eliminating that part of the radiance from the target that is outside the wavelength range to be measured, by placing a cold spectral filter over the detector.

All common sources of noise are random, so are reduced by integrating the signal, that is, sampling it for a long time $\Delta t$ and taking the average. This has no effect on the signal, assuming it remains constant, but the random noise is reduced by a factor proportional to $1/\sqrt{\Delta t}$. The scene has to remain constant during the integration, so $\Delta t$ cannot be very large on a spacecraft moving at the orbital velocity of about 7 km s$^{-1}$. Dwell times of a second or so are the longest that can usually be contemplated.

### 9.5.3   Signal to noise ratio

Combining all sources of noise on a particular measurement leads to a quantity called the *Noise Equivalent Power* or NEP, which has units of W Hz$^{-1/2}$, or W s$^{1/2}$. This parameter characterizes the performance of a given detector and can be measured; a detector supplied by a manufacturer will generally come with a certificate stating the NEP of the device. The definition of NEP is that it is the power in watts that will produce a signal equal to the rms noise for a dwell time of 1 s (so the signal is twice the noise after integration for 4 s, and so on). Obviously, NEP is smaller for better detectors. It varies with the square root of the area $A_d$ of the detector,[101] so a 'quality factor' or *specific detectivity* $D^*$(m W$^{-1}$ Hz$^{1/2}$) can defined by

$$\mathrm{NEP} = \frac{\sqrt{A_d}}{D^*}$$

to give a parameter that is characteristic of that *type* of detector.

We also define the NER or *noise equivalent radiance* of the whole instrument. This is the *change* in target radiance that changes the signal by an amount equal to the noise (in radiance units) for the system as a whole. Implicit in this definition is a value for the dwell time $\Delta t$; unless otherwise specified this is taken to be 1 s($\Delta f = 1$ Hz). The units of NER are therefore those of radiance per $\sqrt{\mathrm{Hz}}$ or W m$^{-2}$ sr$^{-1}$ μm$^{-1}$ s$^{1/2}$.

Related to NER is the NET (sometimes written NE$\Delta T$), which is the change in target *temperature* that produces a signal change equal to the noise. If the instrument is viewing a target with equivalent blackbody temperature $T_B$(K)[102] then it follows from the definitions that

$$\frac{\partial B(\lambda, T_B)}{\partial T}\mathrm{NET} = \mathrm{NER}.$$

While NEP quantifies detector performance and NER or NET does the same for an instrument, the quantity we really want to know in order to evaluate a particular *measurement* is the signal to noise ratio, written S/N or SNR. The signal $S$ when radiance $B(T_B)$ is integrated for time $\Delta t$ by an instrument with etendue $A\Omega$ and spectral bandwidth $\Delta\nu$, is

$$S = gB(T_B)A\Omega\Delta\nu\Delta t,$$

where again the constants of the system have been gathered into a single constant term $g$, the gain, and the offset is assumed to have been removed

[101] See this by considering two identical detectors measuring the same signal. Together they have twice the area, and if their outputs are averaged together the resulting signal has half the noise (for random noise).

[102] In the experimental and engineering literature about remote sensing instruments, and in infrared astronomy, the term *brightness temperature* is often used for the equivalent blackbody temperature of the scene or target being viewed, at the wavelength of the measurement. Brightness temperature is a function of wavelength, as well as of the real physical temperatures of the objects in the field of view. If the target is a blackbody, then of course brightness temperature and physical temperature are the same at all wavelengths.

by calibration. In the same way, the integrated noise $N$ is just

$$N = gNEP\sqrt{\Delta t}$$

so the signal to noise ratio is

$$S/N = \frac{B(T_B)}{NEP} A\Omega\Delta\nu\sqrt{\Delta t}.$$

The NER is, by definition, the value of the radiance when $S/N$ is unity, so

$$NER = \frac{NEP}{A\Omega\Delta\nu\sqrt{\Delta t}}$$

and the NET, also by definition, is

$$NET = \frac{NER}{(dB(T_B)/dT)}$$

and

$$S/N = \frac{B(T_B)\sqrt{\Delta t}}{NER}.$$

where $B(T_B)$ is the radiance from the target with brightness temperature $T_B$.

## 9.6 Limb viewing instruments

Modern atmospheric sounding instruments for the stratosphere and above often use limb scanning, meaning that instead of viewing downwards towards the surface, they look sideways at the limb of the Earth and observe the atmosphere through a telescope (Fig. 9.5). If the instrument has a sufficiently narrow field-of-view, a thin 'slice' of atmosphere is viewed at the tangent point so that the weighting function is much narrower than for

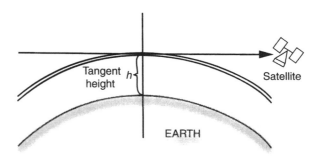

**Fig. 9.5**

The geometry of satellite limb sounding observations. The radiometer on the satellite views in a narrow field-of-view at a tangent height h above the surface of the Earth. The radiance measured is the integral of the emission along the line of sight, with the largest contribution coming from the lowest (highest density) layer at tangent height $h$, but all layers above $z = h$ contribute. By scanning from $z = 0$ to $\infty$, or by using multiple detectors viewing different levels simultaneously, a profile is built up.

vertical or 'nadir' viewing, as can be seen by comparing the limb weighting functions in Fig. 9.6 with those in Fig. 6.12.

The limb weighting functions cut off sharply at the bottom, since the lower atmosphere is not viewed at all. The upper part still extends over a range of heights, due to the fact that the instrument must look through the layers that overlie the one of interest, and they will then make a contribution to the radiance observed in that channel. However, over the height range where the atmosphere is optically thin, most of the radiation originates in a layer that is only 1–2 km thick, which compares very favourably with the 10–15 km wide weighting functions that apply to nadir sounding.

Limb sounding has several disadvantages that offset the large gain in vertical resolution:

1. The horizontal resolution is poorer, in the direction along the line-of-sight (but not perpendicular to that), generally around 200 km. Often, this is an acceptable sacrifice since the most interesting features are those that vary over relatively short distances in the vertical.
2. The instrument design is considerably more complicated, primarily to obtain good optical quality in the form of a narrow, sharp field-of-view without extensive wings or stray light problems, since the temperature gradient in the vicinity of the target is very large in the vertical direction, with cold space on one side and the relatively hot solid planet on the other.
3. Limb-scanning instruments tend to be large, because the narrow field-of-view needed to select a 1 km slab of atmosphere from a distance which is typically about 3,000 km, dictates the need to have quite a big aperture in order to obtain an adequate energy grasp, and to avoid degrading the resolution due to diffraction (Section 9.4.3).
4. The lowest part of the atmosphere (below about 6 km) is difficult to observe, because clouds and aerosol turbidity interfere.

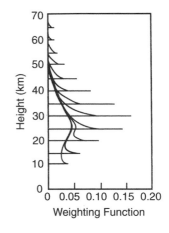

**Fig. 9.6**
Limb-viewing weighting functions, corresponding to a single channel of an instrument physically scanned in the vertical direction in steps of 5 km at a time.

## 9.7  Contemporary satellites and instruments: Three examples

### 9.7.1  Weather satellites: The Geostationary Operational Environmental Satellite

Geostationary Operational Environment Satellite (GOES) is an example of a meteorological satellite equipped with an infrared sounder (Fig. 9.7) to

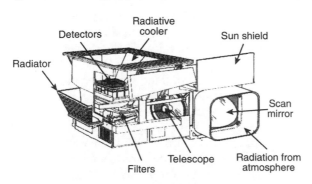

**Fig. 9.7**
An example of an infrared sounder on a modern weather satellite: the Geostationary Operational Environmental Satellite (GOES). The positions of the main components, basically those discussed in section 9.2, are shown. The radiative cooler is a blackened cone, pointing at cold space, which reduces the temperature of the detectors to around 80K.

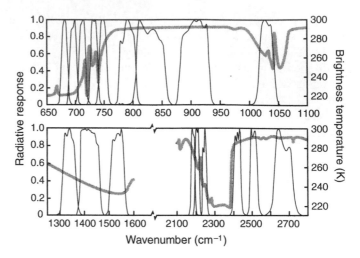

**Fig. 9.8**
The spectral channels used by the GOES weather satellite are shown as the lighter lines; they are expressed as the profiles of transmission versus wavenumber for the spectral filters that define each channel, on a scale of relative response (left) that is normalized to make the maximum transmission equal to one on this scale. The darker lines are the brightness temperature of the Earth as a function of wavenumber (right-hand scale).

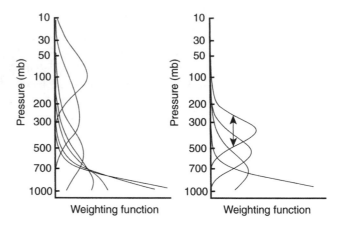

**Fig. 9.9**
Weighting functions, corresponding to some of the channels shown in Fig. 9.8: the longer wavelength channels from 650 to 800 cm$^{-1}$ (about 12.5–15 µm, left) and mid-wavelength channels from 800 to 1 600 cm$^{-1}$ (about 6–12.5 µm, right). The full width at half maximum of a weighting function, illustrated by the double arrow, is an estimate of the vertical resolution of the measurement at that height.

collect data for routine (called 'operational') weather forecasting purposes. Fig. 9.8 shows the wavelength bands (spectral channels) used in the GOES sounder to measure atmospheric (including cloud-top) and surface temperatures, and water vapour profiles. Note the grouping of channels for temperature sounding in the wing of the strong $CO_2$ band near 700 cm$^{-1}$ (top left), comparable to the schematic representation in Fig. 6.10. Other channels cover the major bands of ozone near 1050 cm$^{-1}$ and water vapour near 2,300 cm$^{-1}$, and the spectral 'windows' at 800–1,000 cm$^{-1}$ and 2,400–2,800 cm$^{-1}$.

Figure 9.9 shows the corresponding weighting functions. Any detailed structure with a vertical scale that is less than the width of the weighting functions (usually measured at the half-power point, as shown by the arrows in Fig. 9.9) is averaged out, limiting the vertical resolution of the profile that can be retrieved. This width is typically more than a scale height, so important detail, for example about wave motions and other interesting disturbances in the atmosphere, is often not detected.

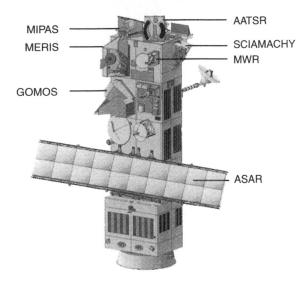

MIPAS — AATSR
MERIS — SCIAMACHY
— MWR
GOMOS —
— ASAR

**Fig. 9.10**
The ESA ENVISAT Spacecraft is 10 m long and weighs 2.673 tonnes. The various acronyms for the payload instruments are; Medium Resolution Imaging Spectrometer (MERIS); Michelson Interferometric Passive Atmospheric Sounder (MIPAS), Global Ozone Monitoring by Occultation of Stars (GOMOS), Scanning Imaging Absorption Spectrometer for Atmospheric Cartography (SCIAMACHY), Advanced Synthetic Aperture Radar (ASAR), and the Advanced Along Track Scanning Radiometer (AATSR), Microwave Radiometer (MWR).

### 9.7.2 Environmental satellites: ENVISAT

The European Space Agency launched ENVISAT, a large polar-orbiting satellite (Fig. 9.10) to provide measurements of the atmosphere, ocean, land, and ice over a 5-yr period beginning on 1 March 2002. Its objectives relevant to climate include quantitative monitoring of ocean–atmosphere heat and momentum exchange, sea surface temperature, atmospheric composition and associated chemical processes, ocean dynamics and variability, atmospheric temperature, water vapour, and cloud top height and coverage.

### 9.7.3 Research satellites: UARS and EOS

UARS is the Upper Atmosphere Research Satellite and EOS is the Earth Observing System. Both are part of a massive programme of NASA in the USA to understand the atmosphere, the weather, including extreme events, and the climate. They carry state-of-the-art remote sounding instruments for climate research, like ISAMS (Improved Stratospheric And Mesospheric Sounder, an infrared limb-viewing radiometer) and MLS (Microwave Limb Sounder) instruments on UARS. These sensors, with dimensions measured in metres and masses of hundreds of kilograms, dwarf the shoebox-sized instruments of the early years of remote sensing in the 1960s. The MLS operates at frequencies in the GHz range, and requires an antenna 1.5 m across to avoid the diffraction limit on the beam width necessary to resolve the limb. At the shorter infrared wavelengths used by ISAMS, only a 25 cm diameter mirror is required, along with cooled detectors and optics to get high sensitivity. Cooling to 77K is achieved using miniature mechanical refrigerators of an advanced design to give high efficiency and a long lifetime.[103]

The EOS is NASA's latest series of giant satellites dedicated to climate studies. The first two in the series, called *Terra* and *Aqua* were launched in 2000 and 2002, with *Aura* following in 2004.

[103] Most cooled detectors are developed to work at 77K because it is the temperature of liquid nitrogen, a convenient laboratory cryogen. The criterion for making the detector cold enough, so that thermal excitation of carriers is negligible compared to excitation by the incident photons, is that the temperature $T$ should satisfy $kT \ll h\nu$, where $k$ and $h$ are Boltzmann's and Planck's constants, and $\nu$ is the frequency of the photon.

The Tropospheric Emission Spectrometer (TES) instrument on *Aura* is a Fourier transform spectrometer with mechanically cooled optics and detectors, designed to measure thermal emission from tropospheric minor constituents with high spectral and high spatial resolution. The main focus is on tropospheric ozone, a key constituent in pollution studies for which no good global data exist, but the wide spectral range means dozens of species (including $NO_y$, CO, $SO_2$, and other sulphur compounds) are observed to address the atmospheric chemistry part of the greenhouse puzzle. An advanced feature of TES is that it can scan continuously between the nadir and the limb. The interesting tropopause region is observed with high vertical resolution, high spectral resolution and continuous spectral coverage, allowing problems of stratosphere–troposphere exchange to be addressed. Convective transport across the tropopause is inhibited by the temperature structure (see Section 3.5) and the ways in which pollutants from the surface reach the stratosphere, and stratospheric ozone enters the troposphere, are still very poorly understood.

Alongside TES on *Aura* (Fig. 9.11) is HIRDLS, which aims to elucidate the complex coupling between atmospheric chemistry and dynamics, for instance the role played by of atmospheric waves in transporting and mixing momentum, vorticity, and chemicals. HIRDLS shares with TES the goal of quantifying the sources of tropospheric ozone, and determining the rates at which ozone-depleting halogen compounds are introduced into the stratosphere and eventually removed.

HIRDLS is an infrared limb scanning radiometer similar to ISAMS that measures temperatures, concentrations of $O_3$, $H_2O$, $CH_4$, $N_2O$, $NO_2$, $HNO_3$, $N_2O_5$, $ClONO_2$, $CFCl_3$, $CF_2Cl_2$, aerosol amounts, and locations of polar stratospheric clouds. Its big step forward is in horizontal and vertical resolution, getting closer to the scale of dynamical activity in the atmosphere, including waves and turbulence, and to better match the grid

**Fig. 9.11**
The NASA *EOS-Aura* spacecraft during final assembly. The Microwave Limb Sounder is at the front, with its oval-shaped dish, then HIRDLS, covered in black insulating material at the centre, then TES, which is the large white box just in front of the mushroom-shaped telemetry antenna.

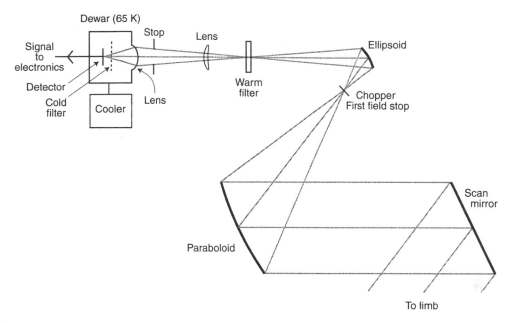

**Fig. 9.12**
The HIRDLS radiometer optics. The radiation from the limb enters at bottom right, and is focussed onto the chopper by the main telescope. After that, the radiation passes through two sets of interference filters before being focussed onto the detector at top left. The detector is cooled for maximum sensitivity, and a second set of filters, just in front of the detector, is also cooled to reduce the background flux detected and so reduce the noise.

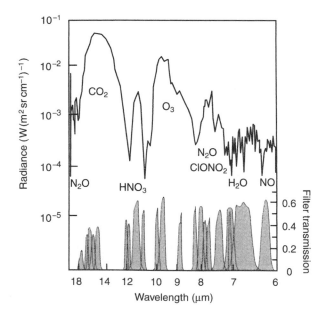

**Fig. 9.13**
The calculated spectrum of the atmosphere, viewed at the limb at a tangent altitude of 20 km. The spectral profiles of the filters that define the 21 channels in the HIRDLS instrument (Fig. 9.12) are shown along the bottom. These are positioned to pick out the spectral bands of the species to be measured, for example, $CO_2$, for temperature, and $ClONO_3$, chlorine nitrate, an important part of the ozone destruction cycle.

spacing used in the best two- and three-dimensional models, like those discussed in Chapter 11.

Figure 9.12 shows the basic optical system of the HIRDLS instrument. This can be thought of a realistic version of the conceptual instrument in

Fig. 9.2, with practical details like fold mirrors (to make the instrument more compact) and stops and baffles (to reduce stray light from the Earth and the Sun when viewing the relatively cold limb). Figure 9.13 shows the filter set for the HIRDLS instrument in relation to the atmospheric spectrum, indicating which atmospheric gases are the measurement objective for each channel.[104] Ideally, the filters would have square profiles and a throughput of 1.00 inside the desired range and 0 outside; these examples represent the current state-of-the art in getting as close to this goal as possible. Notice, as with GOES, the concentration of channels around 15 $\mu$m (600–700 cm$^{-1}$), in order to use the strong carbon dioxide band there for temperature sounding.

[104] Note that the strong absorption bands appear in *emission* (i.e. are the brightest features) in the limb-viewing spectrum, in contrast to a nadir spectrum like that in Fig. 2.9, where they are the coldest features. The reason is that the former is viewing the atmospheric gases against a background of cold space, while the latter is viewing them against a background of relatively warm surface.

## 9.8 Applications of remote sensing to climate studies

Satellite experiments and monitoring using systems like those described above are at the cutting edge of climate research, as, for example, in the following seven areas.

### 9.8.1 Earth's radiation budget

Since the 1960s satellites have measured the reflected solar radiation, that is, the albedo, and the outgoing thermally emitted radiation, of the Earth from orbit. Since the heating by the Sun and the cooling by the Earth are separated in wavelength, falling to the short and long wave sides, respectively, of about 4 $\mu$m (Fig. 2.10), a remote sensing experiment with only the minimal spectral resolution required to separate these can observe and monitor the energy budget of the planet, on both a global and regional basis.

This is a less simple task than it may at first appear, since:

1. The absolute calibration of accurate radiometers is difficult in the hostile environment of space where the gradual deterioration of reference targets cannot be checked.
2. No real instrument can obtain a *uniform* spectral response over the wide wavelength ranges (roughly 0.4–4.0 $\mu$m for solar and 4.0–100 $\mu$m for thermal fluxes) required to measure local energy balance. If the radiation from the target changes its distribution with wavelength, as well as its overall intensity, these are hard to separate when analysing the data.
3. Albedo, although a simple concept, in practice is difficult to derive from data because integration over a $2\pi$ solid angle is required of a field which may have strong directional components (the thermal flux is easier because in this case it is reasonable to expect that cylindrical symmetry applies).

The first successful meteorological instrument on an orbiting satellite was of this kind. Explorer 7, launched on 13 October 1959 carried a simple radiometer with five hemispherical thermistor[105] detectors and crude wavelength selection. Two of the hemispheres had black coatings and responded about equally to solar and terrestrial radiation. A third

[105] A thermistor is a calibrated, temperature-sensitive resistor. See Section 9.4.5.

hemisphere was coated white to bias its sensitivity towards long-wave infrared radiation, while a fourth had a gold covered surface to achieve the opposite effect. With continuous data as the satellite circled the Earth at an inclination of 50° to the equator, it was possible to see for the first time how the energy from the Sun that is absorbed by the Earth is redistributed within the climate system, and finally re-emerges as long-wave infrared radiation.

The individual components of the radiation budget vary on a wide range of distance and time scales, in addition to the dependence on latitude because of mean solar zenith angle and the compensating transport of heat polewards by the atmosphere and the oceans. Even this transport has a seasonal dependence that is not the same every year. Climate models that attempt medium-to-long term forecasts and assess the likely progress of global warming need data on these so they have the correct fluxes of reflected solar and emitted thermal radiation at the upper boundary, a necessary condition for realistic predictions.

More modern satellite instruments address the measurement difficulties by

(1) using redundant standard radiance targets and comparing them to each other to check for changes,
(2) breaking the wavelength range into segments which are measured and calibrated separately, since it is impossible to make the spectral response completely independent of wavelength over a wide range, and
(3) using angular scans which later can be integrated into hemispherical fluxes. The angular coverage over a given region from a single satellite obviously can never be complete, so it is necessary to fit the data to empirical models of the reflectance properties of different regions of the Earth and integrate the model, with a corresponding addition to the error budget.

The Earth Radiation Budget Experiment (ERBE) scanners of NASA have three broad spectral channels, covering the wavelength ranges from 0.2 to 5.0 μm (i.e. effectively the entire solar spectrum), 5–50 μm (most of the Earth's thermal infrared spectrum) and 0.2–50 μm (both). ERBE sensors flew on two sun-synchronous polar orbiting satellites (NOAA-9 and NOAA-10) and on the Earth Radiation Budget Satellite (ERBS). The first of an improved version known as CERES (Clouds and the Earth's Radiant Energy System) was launched into a 35° inclination orbit in 1998 and the second in 2000 on Terra, part of the EOS series of satellites (Section 9.7.3). In Europe, radiation budget sensors have been included on the geostationary platform Meteosat, using a field of view which covers the same part of the Earth at all times of day.

The error in the individual measurements of each component of the radiation budget in a single direction, place, and time has been estimated at about $\pm 5\,\mathrm{W\,m^{-2}}$, or about 2%, but the uncertainty in the overall balance of the planet is considerably larger than this, perhaps $\pm 10\%$ at best. By comparison, the so-called cloud radiative forcing, a useful quantity in climate studies, that is, defined as the flux averaged over all conditions minus the clear-sky flux, is typically about 10–20% of the total. Thus, useful measurements of the *local* energy balance are achievable but *global* measurements with adequate precision are still beyond the current state of the art, especially since our most important goal is to detect and understand changes in the global energy balance.

### 9.8.2 Atmospheric temperature sounding

In section 6.9 we saw that the vertical spread in the source of the radiation measured by an orbiting radiometer is described by the derivative with respect to height of the atmospheric transmission function. This weighting function can be calculated, leaving the temperature of each layer as the principal unknown. The retrieval of the temperature profile then involves inverting the radiative transfer equation, which can be done by a simple iterative method (see Section 6.10), although in practice retrieval methods often include a number of refinements. For example, instead of finding the profile from the measurements alone, statistical a priori information about the expected value of the profile, based on a history of previously measured values (climatology), can be folded in. One way to do this is to use the climatological value of the temperature profile as the starting point for the iteration in a retrieval.

Figure 9.14 shows examples of profiles of atmospheric temperature and humidity that have been retrieved from nadir-viewing radiances measured by an orbiting radiometer. The satellite profiles compare well with ground-based and *in situ* profiles of the same quantity, but have lower vertical resolution, so that any detail on a height scale of less than about 10 km is smoothed out.

**Fig. 9.14**
Temperature and water vapour (expressed as dewpoint) profiles retrieved from a downward viewing satellite radiometer similar to GOES. The significance of dewpoint temperature is that, the nearer it is to the temperature curve, the nearer the atmosphere is to saturation. In this example the troposphere has a high humidity while the stratosphere is very dry, both common conditions.

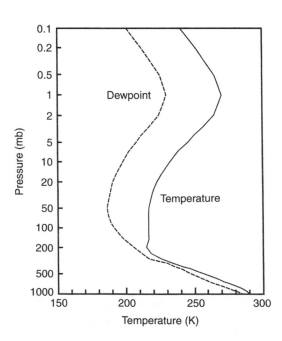

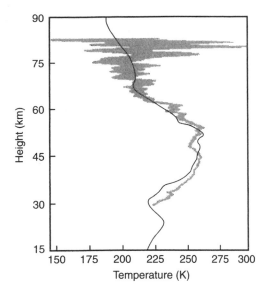

**Fig. 9.15**
A temperature profile retrieved from ISAMS infrared radiances, measured in the $\nu_2$ band of $CO_2$ near 15 μm wavelength at the limb, compared with one measured at approximately the same time and place by a ground-based lidar. The shaded area is the lidar profile, with errors, and the ISAMS profile is the relatively smooth solid line.

Figure 9.15 shows another example of a retrieved temperature profile, this time from satellite limb radiance measurements, compared to one measured at approximately the same time and place by a ground-based, upward-looking optical radar (lidar). Lidar sounding operates on the principle of measuring refractive index profiles, which can be related to density and therefore temperature in the analysis. The resulting profiles have higher vertical resolution (around 100 m, compared to about 1.5 km for a limb sounder, and about 15 km for a downward-viewing radiometer), and more easily reveal the presence of waves and other interesting structure. On the other hand, they are obviously rooted to one spot, and cannot reveal the global extent of wave activity, or the full characteristics in space and time of weather systems, as fully as satellite sensors can.

Satellite instruments produce a profile every few seconds, which can be combined into 3-d temperature maps for use in weather forecasting and climate research. Forecasting involves assimilating information on the current state of the atmosphere into models and then solving the time-dependent equations to obtain the predicted state some days hence. The increasingly sophisticated coverage and resolution of the data competes with the complexity and inherently chaotic nature of the atmosphere until, for forecasts more than about six days ahead, the inherent limits to predictability itself becomes the dominant problem—the climate has its own version of the uncertainty principle.

### 9.8.3 Atmospheric composition and chemistry

Infrared and microwave instruments also measure atmospheric composition. The intensity of the emitted thermal flux depends on the temperature and the abundance of the emitter (i.e. on the probability per molecule of emission of a photon of given wavelength, and on the number of emitting

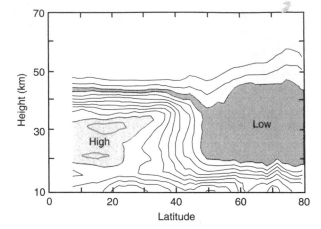

**Fig. 9.16**
ISAMS measurements of nitrogen dioxide, along an orbit track inclined at 23° to the pole. The very large horizontal gradient at 40° latitude marks the transition from night to day. In daylight, most of the $NO_2$ is photochemically converted to nitric oxide, NO.

molecules present). So, once the temperature has been determined from the emission by a molecule of known abundance such as $CO_2$, measurements of the same atmospheric path at a different wavelength in an emission band of a different species can yield information on the concentration of the second species.

Among the first targets were the oxides of nitrogen, $N_2O$, NO, $NO_2$, and $N_2O_5$, following the discovery that they are involved in the catalytic destruction of ozone, posing a threat to the Earth's shield against solar UV radiation. Most members of the $NO_x$ family are very reactive and the amounts available to interact with ozone vary dramatically with space and time and cannot be followed effectively by traditional measurements from aircraft or balloons. Global measurements of nitrogen dioxide reveal large gradients in all three spatial directions (vertically, diurnally, and with latitude, see Fig. 9.16), and nitric oxide in the upper straosphere shows a strong correlation with solar activity. $NO_2$ is removed by conversion to 'reservoir' species such as dinitrogen pentoxide and nitric acid in the long, dark polar night. An understanding of the chemistry going on in all these regions is crucial to predicting the fate of the ozone layer at all latitudes.

The lower atmosphere is experiencing a gradual change due to the accumulation of minor constituents such as carbon dioxide, carbon monoxide, chlorofluorocarbons, nitrous oxide, and methane (Chapter 7). These are radiatively active gases, which affect the energy balance at the surface, in part by radiating back absorbed radiative energy (the greenhouse effect), leading to a general warming. More reactive gases such as ozone are also changing (Chapter 8), and have a profound effect on the chemical mix, which affects not only the 'greenhouse' but also leads to phenomena such as acid rain. The first remote sensing experiments to measure tropospheric carbon monoxide globally for the first time in 1976 found high concentrations, not over the urban industrial centres of N. America and Europe, as expected, but downwind of the forests of S. America and Africa. Since then, the importance of biomass burning has been in the headlines. Today, the EOS and ENVISAT class of satellite is monitoring a wide range of greenhouse gases, ozone destruction precursors, and other pollutants in a global check on the health of the planet.

### 9.8.4   Clouds, aerosols, and polar stratospheric clouds

In addition to their role in radiative and latent heat transfer, precipitation and the water budget, clouds interfere with remote sensing of the lower atmosphere and surface from space and make the retrieval of information from these levels difficult. For some measurements, the problem can be avoided by working at microwave and radar wavelengths, at which most clouds are transparent. Often, however, either this approach does not work for the data we need (spectroscopic observations of composition, for example, mostly have to be made in the infrared where their absorption bands lie), or else we want to obtain remotely sensed data on the cloud properties themselves.

Satellite data has already provided the first reliable information on such basic climate variables as the mean cloud cover of the Earth (more than 60%, or 10% more than earlier estimates). Oceans are cloudier than continents, at 67%, and clouds over the ocean are lower and about 10% less reflective than those over land. The tropics and the temperate zones are generally cloudier than higher latitudes, but the latter are typically almost twice as reflective.

Their effect on the climate depends not only on the amount and distribution of clouds, but also on their microstructure and optical properties. In principle, particle size and composition can be retrieved from multispectral observations, and by studying the reflection and emission properties of clouds at different angles. Tropospheric clouds are mostly too inhomogeneous for much progress to have been made with this so far, but good results have been obtained on stratospheric aerosols. These are sulphuric acid droplets, mostly of anthropogenic and volcanic origin, which are dense enough to affect the radiative balance of the Earth significantly. They also provide surfaces on which heterogeneous chemical reactions can occur, further complicating the ozone depletion problem. We saw in Section 8.7 that the Antarctic ozone hole is caused by a special kind of cloud called a polar stratospheric cloud (PSC) which forms only at very low temperatures (below about 195K) and which contains frozen nitric acid (Fig. 9.17).

The PSCs appear also over the northern polar regions, where their effect on the ozone layer is potentially more harmful to humans; such an event over Scotland and the North Sea was observed in thermal emission at 12.1 $\mu$m wavelength by the ISAMS instrument. A huge tongue of ozone-threatening chlorine oxide was observed at the same time at 1.5 mm wavelength by the UARS Microwave Limb Sounder. In this case, the event was sufficiently localized and short-lived that the net effect on the ozone layer was relatively small, but the phenomenon is obviously of practical as well as scientific interest, especially if pollution continues to increase, and mean temperatures to fall, in the stratosphere, making such events more frequent and of longer duration.

### 9.8.5   Doppler wind measurements

The movement of the atmosphere can be calculated approximately from measurements of temperature gradients, using the primitive equations of fluid dynamics, or by tracking clouds and other tracers such as water vapour.

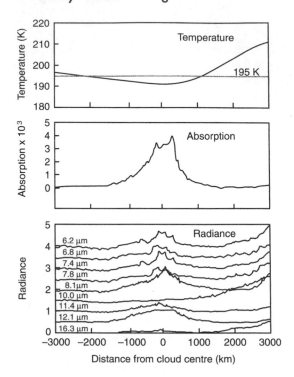

**Fig. 9.17**
ISAMS observations of a polar stratospheric cloud, at a height of about 20 km (46 mb). The central frame shows the absorption profile of the cloud at the 'window' wavelength of 12.1 μm, at which all of the opacity is due to the cloud. Below are radiances at the various wavelengths; analysis of these revealed the composition of the cloud to be $HNO_3 \cdot 3H_2O$. The uppermost curve shows the temperature measured at the same locations, showing that the cloud is only present when $T < 195K$, the freezing point of nitric acid.

Direct wind measurements are also becoming possible by measuring the Doppler shift of spectral lines emitted from atmospheric gases, although at present the technique is only applicable to the middle and upper atmosphere where the lines are narrow, and easier to observe at the limb. Accuracies of a few m s$^{-1}$ are needed from a spacecraft, which itself is travelling, relative to the surface, at something like 7 km s$^{-1}$, mandating a very stable spacecraft with accurate pointing. A simple calculation shows that one degree of uncertainty in the pointing vector of the instrument introduces an uncertainty of around 50 m s$^{-1}$ in the line of sight velocity, about the same as the mean magnitude of stratospheric winds. To obtain a useful accuracy of 1 m s$^{-1}$, pointing knowledge of around an arc minute is required. The Nimbus meteorological satellites of the 1960s and 1970s wobbled in their orbits by several degrees, while modern platforms like UARS are orders of magnitude better and have demonstrated useful wind measurements with interferometric instruments, based on the principle of the Fabry-Perot etalon (Section 9.4.4).

### 9.8.6 Surface properties

Remote sensing is a multi-billion pound industry in which land usage, cartography, surveillance, and crop monitoring, to name but a few, are increasingly feeling the benefits. For instance, the world's rice crop, which is a stable food for more than two billion people, can be followed through its growth and reproduction cycle by mapping from space using radar. Radar penetrates cloud cover and satellite monitoring promises to replace the present highly erratic methods that are used to assess acreage and expected seasonal yield. The signals that correspond to different states of the crop are learned by comparison with 'ground truth', rather than from fundamental principles, although quasi-physical modelling is a growing research field.

Compared to the atmosphere, radiometry and spectroscopy of the surface of the Earth from space is relatively easy, because the target is warm and bright, providing strong signals. The interpretation of the signals, however, is harder. Surface temperature measurement is simplest over the ocean, because the infrared emissivity is close to unity. However, the surface temperature contrasts being sought are quite subtle, measured in fractions of a K, and can be masked by atmospheric effects, for example, very thin clouds or patches of aerosol, which can be quite undetectable in the visible. The Along-Track Scanning Radiometer on Envisat (see Fig. 9.10) overcomes this problem by viewing the surface at two angles and analysing the difference between those views, which would be nearly zero if the atmosphere had no effect.

When observing the land, one is more likely to assume a temperature from other information and examine the observed differences in emissivity. By day, the reflectance properties at visible and near-infrared wavelengths show differences such as that between vegetation and bare soil, and careful choice of wavelength filters (e.g. isolating the chlorophyll bands) enhances sensitivity to vegetation types and different stages of growth. Other interesting possibilities use advanced spectral methods to address the mineralogy and physical state of the surface, with a view to studying desertification, for example, or even prospecting for oil or other valuable deposits. The problem mainly is the contaminating and obscuring effects of the atmosphere, the relatively bland spectra of most minerals, and the fact that material of different kinds is always mixed and layered in any scene. Spectroscopic data are sensitive to so many factors, including relatively uninteresting parameters such as fractional cover, shadowing and dampness, that detecting and classifying change is often easier than identifying the basic cause. Consequently, although remote sensing of the Earth's land surface is a major industry, the techniques are often more a matter of empirical models, and acquired skills such as image processing, than basic physics. Remote sensing of land surfaces is important where no alternative is available, for example on planetary bodies, especially relatively airless ones like the Moon, Mercury, Mars, and the giant satellites of the outer planets.

During the course of the year, the snow and ice cover of the northern hemisphere varies by more than a factor of two. These changes are difficult to monitor from the ground, but can be observed from space, and microwave techniques allow not only the coverage but also the depth of the snow field to be obtained. The dielectric constants of snow, water, and ice are so different that the effect on microwave emission is easily detected. Dry snow attenuates the emission from the surface underneath, mainly by scattering rather than absorption, and allows the depth of the layer to be estimated. Over the ocean, the coverage of sea ice can be mapped by making use of the contrast between the emissivities of ice and water (at a wavelength of 1.55 cm, for example, the emissivity of sea water is 0.44, whereas that of ice is approximately twice this).

The elevation of the ice, hence an estimate of its thickness, is available from space-borne radars with theoretical accuracies of around 1 cm, although the variance in the terrain within the footprint limits the range precision to around 30 cm for smooth but undulating ice sheets, and a few metres over rough terrain. Improvements result from the use of laser altimeters, which have a much smaller footprint, especially if retro-reflectors can be placed at key points on the surface, for example, on an icesheet or glacier whose thickness and movement is to be followed. The precision in that case can be centimetres or even millimetres. Devices of this type have already been flown to the Moon and Mars, and on NASA's Earth Observing System of satellites.

### 9.8.7 Detection of climate change

Satellite measurements of temperature, albedo, ice cover, ozone, and all the other 'fingerprints' of climate variability offer the best way of detecting and monitoring global change. However, even such simple concepts as the global mean surface temperature are notoriously difficult to relate to actual measurements, even from space where extensive (if not complete) and consistent (if not necessarily accurate) coverage with a single instrument, or set of similar instruments, is possible. Microwave data might be particularly suitable for monitoring temperature trends, because the flux is insensitive to the presence of clouds, which are the most variable feature. However, it actually poses a puzzle.

Eight versions of an instrument called the Microwave Sounding Unit or MSU have flown on the TIROS and NOAA series of operational satellites since late 1978, and from the data temperature trends for the lower troposphere and the lower stratosphere have been produced. Interestingly, the trend during the last 20 yrs is an insignificant 0.05K/decade (Fig. 9.18), much smaller than the prediction of around 0.3K/decade from the best climate models and from the aggregate of surface measurements to which the models are most frequently compared (e.g. Fig. 11.1). Several reasons have been advanced for this. One, preferred by the MSU experimenters

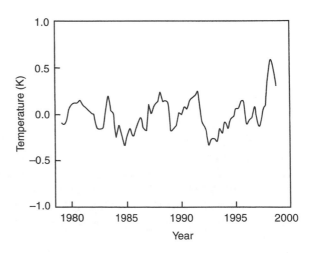

**Fig. 9.18**
Global temperature trends in the middle and lower troposphere over the 20-yr period 1979–1998, as determined from the Microwave Sounding Units mounted on various weather satellites. (NASA Global Hydrology and Climate Center).

themselves, is to accept that there is a genuine difference between the temperature at the surface and in the free air in the troposphere. They invoke a complex set of processes, not fully understood, involving convection and latent heat transfer to account for this. In support of a 'meteorological' explanation of this kind, they note that the warming trends detected by MSU vary considerably over the globe, and are often negative, especially in the southern hemisphere. These trends tend to cancel out when a global mean is taken.

Others argue that the MSU measurements are prone to large errors, pointing to the very large vertical spread of the weighting functions for microwave measurements in the troposphere. Any single channel measures radiation from not only the lower troposphere, but the upper troposphere, the surface, and part of the stratosphere as well. These unwanted contributions have been carefully subtracted out to obtain the trend shown in Fig. 9.18, but the error analysis contains uncertainties that may be of the same order as the effect being studied. Finally, there are those who find in the MSU data convincing evidence that there has been no significant global warming in the last 20 yrs.

## 9.9 The future

Remote sensing is a young field and is continually being refined and improved. Larger, more stable spacecraft offer platforms for bigger, more sensitive instruments that can map increasingly finer detail on the surface and in the atmosphere. The next breakthrough is likely to be the further development of *active* remote sensing techniques. Radar is used for studying the height and roughness of the ocean surface (Section 5.2) and lidar to improve surface altimetry and to obtain Doppler wind measurements. The use of space-based radars to detect and map bands of precipitating clouds is likely to come of age soon. Most recently, temperature profiling is being attempted using the satellites of the Global Positioning System, designed for navigation, as radio beacons for a receiving satellite, which then retrieves vertical temperature profiles by first inferring the density profile from the refraction of the radio beam.[106]

Looking further forward, occultation techniques using two or more satellites, in which one carries a source (probably again a high-powered laser) and the other a receiver, could in principle revolutionise research requiring temperature and composition measurements by providing very high vertical resolution and spectral sensitivity. Any species that is detectable in the laboratory by spectroscopy might be monitored in the atmosphere in this way, including the more exotic pollutants, present in only trace amounts.[107]

## Further reading

*Observing Global Climate Change*, K. Kondrat'ev, A.P. Cracknell, and K. Ya. Kondratyev, Taylor & Francis, Abingdon, 1997.
*Earthwatch: Climate from Space*, by John E. Harries, Wiley-Praxis Series in Remote Sensing, Chichester, 1994.

[106] Difficulties are experienced because of attenuation by water vapour, especially in the troposphere where meteorologists most need data. The technique works better on Mars, where it is likely to be implemented before long.

[107] Experiments have already been implemented in which limb-viewing spacecraft use the Sun as a source to obtain high signal to noise ratios using high spectral resolution spectrometers, but of course the measurements are then limited in time to local sunrise and sunset.

*Remote Sensing and Climate Change: The Role of Earth Observation*, Arthur
    P. Cracknell, Springer-Praxis Books in Geophysical Sciences, Springer-Verlag,
    UK, 2001.
*The NASA Earth Observing System*, http://eospso.gsfc.nasa.gov/
*The ESA ENVISAT Programme*, http://envisat.esa.int/

## Questions

1. Discuss the relative merits of satellite-borne and ground-based measurements
   for remote sensing of climate variables, with emphasis on (a) the mean global
   temperature of the Earth's surface, and (b) the solar constant.
2. Explain the principles of atmospheric temperature profile measurement by
   satellite radiometry, using (a) nadir and (b) limb sounding.
3. Radiances $R_1 \ldots R_n$ are measured over a range of absorption coefficients in the
   15 μm $CO_2$ band by a nadir-viewing satellite radiometer. Derive an expression for the vertical temperature profile of the atmosphere in terms of these
   radiances.
4. Draw a block diagram indicating the main components required for an infrared
   radiometer designed to measure temperature as a function of height by viewing
   the limb of the atmosphere from a spacecraft.
      Indicate briefly what factors are involved in choosing (a) the spectral region,
   (b) the angular field-of-view, (c) the entrance aperture, and (d) the calibration
   scheme used by a limb-scanning radiometer.
5. The 'quality factor' $D^*$ of an infrared detector is defined to be $\sqrt{A}/(\text{NEP})$
   where $A$ is the area of the detector. What is meant by the noise equivalent
   power (NEP) of the detector?
6. Derive a formula to express the noise equivalent temperature (NET) of a
   radiometer in terms of its geometry, the spectral bandpass and optics transmittance, the integration time, and the detector noise equivalent power
   (NEP).
7. An infrared radiometer has entrance aperture 0.01 $m^2$, and its bandpass is from
   660 to 670 $cm^{-1}$. The transmittance of the optics is 50% and the field of view
   is 10 mr. Estimate the integration time needed to obtain a NET of 1K if the
   detector NEP is $10^{-9}$ W $(\text{Hz})^{-1/2}$, and the target temperature is 250K.
      [You may assume that the Planck function at $\nu = 665$ $cm^{-1}$ and temp-
   erature $T$ K is given by the following approximation: $B(\nu, T) = 2 \times
   10^{-11} T^4 \, m^{-2} \, sr^{-2} (cm^{-1})^{-1}$].
8. A radiometer is deployed on an Earth satellite in order to investigate the vertical
   distribution of temperature and minor constituents in the atmosphere. If global
   measurements at the highest possible vertical resolution are required, what
   orbit and observing geometry would be appropriate? What quantities would
   be measured by the instrument, and how are these related to the atmospheric
   properties of interest? Explain briefly how the latter could be obtained from
   the measurements.
9. The altitude of a satellite orbit is 500 km. Assume that the detector is square
   and has $D^* = 3 \times 10^8$ m $Hz^{1/2}$ $W^{-1}$ and that a spectral filter of width 10 $cm^{-1}$
   at 1 000 $cm^{-1}$ is used. [The radiance emitted by a black body at 1 000 $cm^{-1}$ is
   38 m W $m^{-2}$ $sr^{-1}$ $(cm^{-1})^{-1}$].

   (a) What angular field of view is required to obtain a geometric resolution of
       1 km in the vertical direction?
   (b) What would be the minimum size of the main telescope mirror required to
       avoid degradation of this geometric resolution?
   (c) Calculate the physical size of the detector consistent with (a) and (b).

(d) Calculate the signal integration time for a horizontal resolution of 50 km.
(e) Calculate the signal-to-noise ratio when viewing the atmosphere at a height where the optical depth is unity and the temperature is 250K,

10. Comment on the actual vertical and horizontal resolution that can be expected in the retrieved profiles of temperature and composition from measurements with these parameters.

# 10 Climate sensitivity and change

## 10.1 Introduction

The climate on the surface of the Earth is the result of a balance between a large number of factors, all of which can change. If the heat from the Sun increased, for example, we would expect the surface of the Earth to respond by getting hotter. If the amount of carbon dioxide in the atmosphere increases, this also tends to produce an increase in surface temperature. These are examples of external and internal, respectively, climate *forcing*.[108] How sensitive is the climate to different kinds of internal and external forcing? Can we estimate the relative importance of each variable, in terms of its likely effect on the climate?

We can try to answer these questions using models. We can also look at evidence from the past, by analysing the *paleoclimate*.[109] We know that the climate has changed over the ages, sometimes dramatically, but it is not simple to reconstruct exactly what happened and to attribute it to a specific cause. There is evidence that the present day climate is changing but no complete consensus about the nature and extent of the changes, or the timescale on which they are expected. As we have seen, attempts to predict future change, or even to understand past change, are made complicated by the tendency for perturbations to be damped, moderated, and in some cases accelerated, by feedback mechanisms of various kinds.

In this chapter we will review the main factors that are thought to be prime candidates for inducing past, present and future climate change, under the following five headings:

1. Changes in the Earth's orbit, that alter the planet's distance from the Sun, or in its obliquity, the angle between the spin axis and the orbital plane.
2. Changes in the energy output of the Sun.
3. Changes in the albedo of the Earth, affecting the fraction of the solar input that is reflected back into space.
4. Changes in the composition of the atmosphere, altering the greenhouse warming of the surface.
5. Changes in the ocean circulation, affecting the distribution of incoming solar energy around the globe.

There are also various 'one–off' or catastrophic changes that could dramatically change the climate, which we will not consider further here. These include things like the possibility of a collision with the Earth by even a moderately sized asteroid or comet, the debris from which could alter

[108] Note that climate *forcing* does not necessarily lead to significant climate *change*, since more than one forcing may be present, and they may balance each other, either coincidentally or as a result of negative feedback, or both. But forcing does *tend* to produce change.

[109] 'Paleo' is from the Greek palaios meaning 'ancient'.

the composition and cloudiness of the atmosphere very substantially. The resulting changes in the greenhouse effect could render the surface virtually uninhabitable for many years or decades. The theory that such a catastrophe led to the end of the dinosaurs, after millions of years as the dominant species on Earth, is well known. The phenomenon of a planet-wide 'nuclear winter', following a war, could have the same sort of effect on all of those who survive the war itself. A natural version of nuclear winter, involving the simultaneous eruption of a large number of major volcanoes, is perhaps not as probable on the present-day Earth as it may have been in the past, but this is a candidate for the warm, wet epoch on early Mars, and may explain the current situation on Venus.

## 10.2 Astronomical changes

### 10.2.1 The Milankovich cycles

The Earth's orbit is affected by the gravitational attractions of the Sun, the Moon, and the other planets. This complex interaction results in slow, cyclic changes in three important parameters of the orbit:

1. The *eccentricity* of the ellipse that the Earth describes in its orbit around the Sun each year.
2. The angle between the equatorial plane and the orbital plane (the Earth's *obliquity*).
3. The *precession* of the spin axis around the normal to the orbital plane (Fig. 10.1).

The name that has come to be associated with these cycles is that of Milankovich,[110] who showed that the periodic variations in climate deduced

[110] Milutin Milanković (Milankovich or Milankovitch), 1879–1958, was Professor of Applied Mathematics at the University of Belgrade. His work on the effects of orbital variations on climate was collected in 1941 into *Kanon der Erdbestrahlung und seine Anwendung auf das Eiszeitenproblem*, published by the Royal Serbian Academy, and translated into English in 1969 under the title *Canon of Insolation of the Ice-Age Problem*.

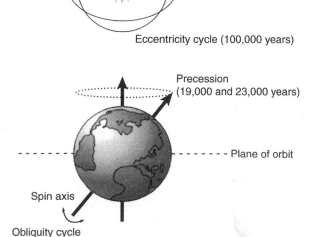

Earth's orbit

Eccentricity cycle (100,000 years)

Precession
(19,000 and 23,000 years)

Plane of orbit

Spin axis

Obliquity cycle
(41,000 years)

**Fig. 10.1**
The three Milankovich cycles, in orbital eccentricity, obliquity, and precession of the spin axis, and their dominant periods.

from glacial records can be matched to the calculated periods of these changes in the orbit of the Earth. These calculations are complex, since the variations are not simple harmonics; many periodicities are present in all three cycles.

[111] Astronomical solutions for paleoclimate studies over the last three million years, A. Berger and M.F. Loutre, *Earth and Planetary Science Letters*, **111**, 369–382, 1992.

Figure 10.2 shows some modern results[111] for a calculation of each of the Milankovich cycles, and their combined effect on the annual mean insolation at one particular location (65°N) on the Earth. Since the mean distance of the Earth from the Sun does not change with any of these cycles, the net insolation averaged over the whole planet does not change significantly; any effect on the climate is through the changes in the seasons. For example, the eccentricity of Earth's orbit varies between about 0.00005, which is nearly circular, and 0.05, corresponding to a variation of as much as 5% in the distance of the planet from the Sun during the year, or roughly 10% in insolation. We are presently at an intermediate value of 0.017, so the solar power per unit area at the Earth varies by about 3% during the year. The effect of orbital eccentricity is to intensify the seasons in one hemisphere

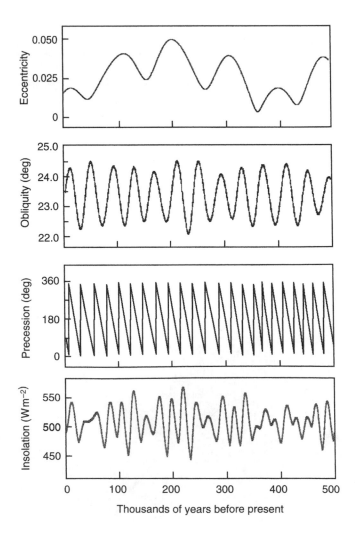

**Fig. 10.2**
Calculations of how the orbital cycles from the previous figure progress with time, and the resulting effect on the solar power per unit area reaching the Earth, on a time scale spanning 500,000 yrs. The insolation is calculated for a latitude of 65°N on 15th June; the eccentricity is the departure of the orbit from a perfect circle, the obliquity is the angle between the spin and orbit angular momentum vectors, and precession refers to the angular position of the Earth in its orbit at the time of the northern summer solstice.

(the one that has summer when closer to the Sun) and moderate them in the other (which has winter when closer to the Sun).

The obliquity presently has a value of 23.5° and is predicted to vary between 22° and 24.5°. If it were zero, there would be no seasons; larger values means warmer summers and cooler winters in both hemispheres, especially at high latitudes. The precession of the spin axis means it slowly circles, like a top, around the upright position, changing the phasing of the seasons relative to the position of the Earth in its elliptical orbit. Currently, midsummer in the northern hemisphere occurs when Earth is farther from to the Sun than it is during southern midsummer, so the south tends to have slightly warmer summers and cooler winters than the north. This situation will reverse after half of a precession cycle (about 10,000 yrs).

The components of the Milankovich cycles that have significant amplitude all have very long periods. In the case of the eccentricity, the dominant cycles have periods of about 100,000 and 400,000 yrs, with the result shown in Fig. 10.2. The dominant periodicity of the obliquity variation is about 40,000 yrs, while the precession of the spin axis is relatively rapid with two large components at about 19,000 and 23,000 yrs. This leads us (tentatively at least) to rule out orbit changes as a cause of climate change on decadal timescales, like that we are experiencing at present.

## 10.2.2  The ice ages

It is well known that numerous advances and retreats of polar glaciation have taken place during the history of the Earth. These events do not seem to have been randomly distributed in time but were concentrated into four time intervals: between about 800 and 600 million years ago, 460 and 430 million years ago, 350 and 250 million years ago, and in the last 3 million years. Nor are they all of equal severity: in major ice ages most of the planet is covered; in minor episodes, like the last ice age which occurred about 15,000 yrs ago, the ice sheet extended down to about 40° latitude, far enough to cover Britain and Canada but leaving most of the global land mass ice-free. We are within the cold phase of a small-amplitude cycle at the present time, but in a short, relatively warm period between glacial advances. Over 20 of these advances and retreats have occurred during the last 2 million years.[112]

There have been both rapid and gradual periods of change. The major ice ages apparently began with millions of years of gradual cooling, accelerating as the albedo of the planet increased due to the growth in the area covered by ice, which is much more reflective than land or ocean. Tracking this kind of behaviour is made more complicated by continental drift: for instance, there is evidence that, when the current ice age started some 3 million years ago, North America, Europe and Asia were at near-polar latitudes and experienced extensive glaciation. Much earlier, parts of Africa, South America, India, and Australia, may have been at the south pole when they were heavily glaciated. A hundred million years ago, the geological evidence shows that Antarctica was covered in vegetation and it became ice-covered only about 35 million years ago.

[112] 'Relatively warm' means warm for an ice age; in the longer term, the Earth has been warmer than it is now and the record as presently understood shows that there was no permanent ice, even at the poles, most of the time during the last billion years.

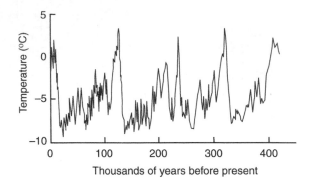

**Fig. 10.3**

Atmospheric temperature over the last 400 millennia as deduced from trapped gases in Antarctic ice cores. When interpreting these, it has to be kept in mind that the variation in air temperatures over the icecaps is not the same as changes in the global mean temperature. Nevertheless, several large temperature swings—the ice ages—are clearly recorded. (Data from National Ice Core Laboratory, US Geological Survey.)

The best data we have on temperatures during the ice age cycles comes from the analysis of air trapped in pockets in the polar ice, although the record is only about half a million years long (Fig. 10.3). With this technique, fairly precise dating is possible from the progressive layering process that laid the ice down. The temperatures are based on measurements of isotopic ratios, especially $^{16}O/^{18}O$ and D/H, since the amount of the rarer isotope found in the ice is a sensitive function of the temperature at which the ice formed.

The data from ice cores does indeed show periodic cooling with quite a good correspondence with the calculated periods of the Milankovich cycles.[113] It is possible to discern with the eye a major period of about 100,000 yrs in Fig. 10.3, which is encouragingly close to the dominant period of the cycle in the eccentricity of the Earth's orbit. A full spectral analysis of this and other data has extracted cycles with periodicities of 43,000, 24,000, and 19,000 yrs, all of which match the astronomical cycles illustrated in Fig. 10.1 quite closely.

All is not rosy for the Milankovich model of the ice ages, however, for at least two basic reasons:

1. The cycles do not increase or decrease the total energy received by the Earth if this is averaged over the course of a year. An increase in eccentricity, or in obliquity, means the insolation is larger during part of the year, and smaller during the rest, with very little net effect on the climate. It has to be assumed, therefore, that the longer-term secular changes in the extent of the ice caps, etc. were forced by changes in the severity of the seasonal cycle. This would involve some quite complicated non-linear effects that are hard to model or explain.

2. Even given the preceding assumption, we then would expect an asymmetric climate change, in which one pole became more ice covered while the other became less so. And yet, the geological record shows that the whole globe cooled during the ice ages. Perhaps the freezing of one

[113] There is also a remarkable correlation between the carbon dioxide and methane abundances in the trapped atmospheric samples, and the global mean temperature. The dating is not precise enough to say whether the atmospheric composition change is the cause of, or a result of, the temperature rise, nor what the drivers were in either case. Presumably the gases were released during the periods of deglaciation from reservoirs in the ocean, the $CO_2$ dissolved in the water and the methane trapped in seafloor sediment.

pole sufficiently suppressed the thermohaline circulation in the ocean that the other pole froze also, despite enjoying increased insolation.

The need for these plausible but sophisticated assumptions means it is difficult to estimate whether the magnitude, and not just the period, of the astronomical cycles matches the observed climate. Attempts have been made using complex models, with some success, but currently the calculated amplitudes fall short of those inferred from the data.

### 10.2.3   Current changes

Setting aside for the moment the remaining uncertainties in our understanding of the paleoclimate, we can consider what role the Milankovich cycles play in the present-day fluctuations in climate that are our primary concern. Here, at least, a simple answer is forthcoming. The key point is that the orbital changes that are large enough to modulate the seasonal cycle are very slow; those terms in the Milankovich cycles that have frequencies short enough to be measured in centuries are very small in amplitude. The best calculations of the Earth's orbit do not show, based on past correlations, another ice age, even of modest proportions, due to this cause, for another 50,000 yrs.

The implication is, therefore, that, while the more distant ice ages could have been caused by this type of mechanism, they are probably immaterial in the context of current concerns about climate change on the timescale of a century or less.

## 10.3   Variations in solar output

Changes in output from the Sun are expected to have a fairly direct effect on the surface temperature of the Earth. In energy balance models (Section 7.4), the effective emitting temperature of the planet is proportional to the fourth root of the solar constant, while the stratospheric temperature $T_s$ is proportional to the Earth's radiative balance temperature $T_E$, and any change in $T_E$ is numerically identical to the accompanying change in surface temperature. Formally:

$$T_s \sim T_E \sim S^{1/4}$$

and

$$\Delta T_E = \Delta T_o,$$

keeping in mind of course that this is a zeroth-order estimate that ignores all feedbacks, and assumes equilibrium despite the long time constants involved.

The radiation from the Sun varies on a range of time scales, from days to millions of years. Accurate measurements of the actual value of the irradiance reaching the Earth have been available only since 1978, when it became possible to make measurements above the atmosphere for the first

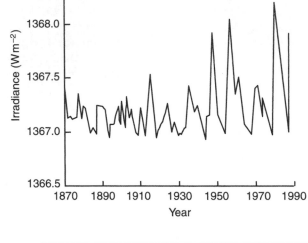

**Fig. 10.4**

Solar irradiance variations during the last century, inferred from proxy data such as sunspot number and cycle length; solar diameter and rotation rate; and the abundance of cosmogenic isotope ratios in tree rings and ice cores, which are indicators of the strength of the solar magnetic field. (Solar irradiance since 1874 revisited, S.K. Solanki and M. Fligge, *Geophysical Research Letters*, **25**, 341–344, 1998.)

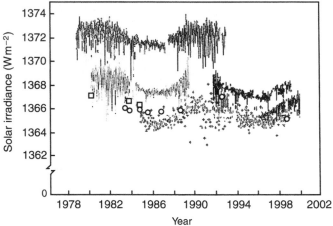

**Fig. 10.5**

Solar iirradiance data from several different instruments on rockets and orbiting spacecraft[114]. Note the rapid variations due to sunspots, and the slower variations with amplitudes of less than 0.1% that are of the same order as the discrepancies between the measurements from different instruments.

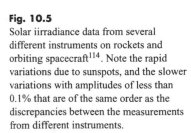

[114] The continuous data sets are from the Earth Radiation Budget instrument on Nimbus 7; the Active Cavity Radiometer Irradiance Monitor on the Solar Monitoring Mission (1980–1990) and on the Upper Atmosphere Research Satellite (1991–1997); and the Earth Radiation Budget Satellite. The discrete points (large squares and circles) are the results from rocket-borne measurements. (NASA).

time. Before that, it was necessary to rely on monitoring 'proxies' such as sunspot number, solar spectral features and the abundances of certain radioactive isotopes in the Earth's atmosphere produced by cosmic rays that are modulated by solar activity. The results are thought to be fairly reliable indicators of the solar irradiance back to about 1874 (Fig. 10.4).

The variation in $S$ is mostly related to the 11-yr cycle in the number and size of sunspots, huge magnetic storms that show up as cooler, and hence darker, areas on the Sun's surface (Section 2.2). The peak-to-peak change in energy radiated during that cycle is less than one part in a thousand, corresponding to a variation in $\Delta T_0$ of about 0.05K according to the energy balance model.

Of more interest are the data for the last four decades, which should be more accurate and which allow us to look for signs that changes in the Sun may be responsible for the recent global warming trend. Figure 10.5 shows some of the latest and best measurements of solar irradiance made from several instruments on satellites. Three features stand out in this:

1. The instruments do not agree precisely with each other. This is due to the problems of obtaining an absolute calibration. Even so, the scatter between measurements is less than 1%.
2. The variation in the measurements from a particular instrument over the 20-yr period is about 0.1%.
3. The 'noise' on the plots is again due to sunspots; these more rapid fluctuations are also of low amplitude, around 0.1%.

Despite the small observed variations in irradiance, some researchers have found correlations between sunspot number and atmospheric temperature, and with related parameters like rainfall and stratospheric ozone abundances. If these are not spurious, it is possible that some poorly understood feedback process in the atmosphere, perhaps involving clouds, is amplifying small changes in the sunfall beyond that presently predicted by models.

Climatically important changes in solar output undoubtedly occur on very long timescales: there is evidence that the solar constant was 2.5% less 300 million years ago, for example, and, like other stars, the Sun will evolve and change its radiating temperature and colour over the eons to come. However, as with the Milankovich cycles, there is no compelling evidence at present that solar fluctuations are responsible for, or expected to produce, climate change on human timescales.

## 10.4  Changes in atmospheric composition

### 10.4.1  Greenhouse gases

We saw in Chapter 6 how the present-day climate is maintained by the greenhouse effect, without which the surface temperature of the Earth would be around 30K cooler on average. The largest single contributor to the greenhouse is water vapour, followed by carbon dioxide, methane, and nitrous oxide. Since the Earth has free water on its surface, the amount of vapour in the atmosphere is determined primarily by the temperature. If the other greenhouse gases were removed, the temperature would fall enough to remove most of the water vapour as well, and the Earth would freeze.

This is not in prospect, however, since in fact the greenhouse gases are all increasing. Carbon dioxide, in particular, has increased by about 20% in the last 50 yrs and methane by even more, around a factor of 2 (see Fig. 1.11). The reason for the methane increase is somewhat obscure. It may be the increased cultivation of rice in paddy fields, where the anaerobic conditions lead to methane production. In the case of carbon dioxide it is fairly obvious, since the gas is produced in combustion processes of all kinds, including those in factories, power stations, and vehicles. Temperature rises forced by these increases will increase the moisture content of the air, and further increase the greenhouse warming.

Given these increases in greenhouse gas concentration, it seems inevitable that the temperature of the planet will rise, as indeed it has (Fig. 10.6).

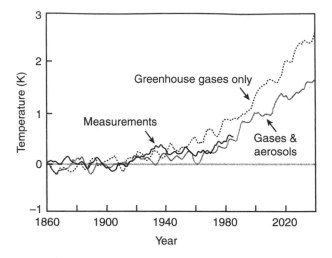

**Fig. 10.6**
Actual mean global temperatures since 1860 (heavy line) compared to model calculations with, and without, the radiative effects of sulphate aerosols. The model calculations have been run out to 2040, and show that aerosol build-up is expected to reduce global warming by about a degree. (Intergovernmental Panel on Climate Change, IPCC.)

However, the rate of the increase depends on the negative as well as positive feedback processes at work. For instance, the expected increase in water vapour could lead to an increase in cloudiness and hence in albedo, tending to bring the temperature back down. Effects of this kind mean the observed increase in temperature is smaller than that predicted by models based solely on the effect of increases in greenhouse gases.

In fact, current research is pointing the finger, not at clouds of the familiar kind but at aerosols, the dispersions of minute airborne particles that form hazes at high, as well as low, levels in the atmosphere (Section 4.10). Like the greenhouse gases, these are on the increase as a result of human activities, but in this case the dominant forcing is towards cooling.

### 10.4.2   Aerosols

Aerosol hazes scatter incoming sunlight, increasing the albedo of the planet, absorb both solar and long-wave radiation, and affect cloud formation and precipitation rates. Generally, their net contribution is to reduce the amount of solar energy reaching the Earth's surface, although it is difficult to quantify these effects precisely. It is much harder to measure the abundance and effect of a highly variable, very fine distribution of tiny sulphate particles in the stratosphere than to monitor the build-up of a well-mixed, stable gas like carbon dioxide.

Nevertheless, since their importance was realized a decade or so ago, computer simulations of climate change have included a primitive representation of aerosols and found that they can offset a significant fraction of the warming due to accumulation of greenhouse gases in the atmosphere. With the usual caveats about the simplifications involved, results like those in Fig. 10.6 show how reasonable assumptions about aerosols can bring model results in line with observations of recent climate trends, and partially ameliorate the doom-laden predictions for the global warming expected by 2,100 (although they still remain serious in the Intergovernmental Panel on Climate Change (IPCC) predictions, which include the expected contribution of aerosols). Another effect of these model findings, combined with theoretical expectations and field observations of aerosols and their impact, has been to place research on the physics of aerosols firmly on the agenda in efforts to improve climate forecasts.

### 10.4.3  Cloud and albedo variations

We have seen that the albedo of the Earth is largely determined by the cloud cover, and that even small trends towards long-term changes in the amount or type of cloud can be expected to have a serious effect on climate. The simple energy balance model of Chapter 7 predicts a change in the temperature of the Earth of about 1K for each change of 0.01 in the albedo, which presently has a value of around 0.30. Since a change of 1K or more in global mean temperature would certainly constitute significant climate change, and since our knowledge of the albedo at any moment is certainly poorer than 0.01, the possibility that cloud cover is a crucial climate variable becomes evident.

The key question is whether the fact that clouds vary locally a great deal means that the net effect of clouds on climate, averaged over a season or longer, is not constant. Day-to-day, and place-to-place fluctuations must average out to a considerable degree, but we cannot be sure that there is not some long-term drift in the mean effect. This is extremely difficult to study, since, as we have seen, it is not just cloud amount that affects climate but also factors such as cloud thickness, microphysics, and height in the atmosphere, since these change the albedo and the internal energy balance of the system.

Global monitoring of these factors is patchy at present, as is our understanding of processes that might affect cloud behaviour. Several speculations have already been mentioned:

(1) small changes in solar radiation (particles, as well as photons) might change the amount of cloud and hence the planetary albedo in such a way as to amplify the effect of changing sunspot numbers on the surface temperature on the Earth, and

(2) anthropogenic pollutants, especially aerosols, affect the annual average amount and type of cloud present, plus, once again

(3) global warming itself induces feedback involving clouds, since their formation and dissipation processes are temperature-dependent.

Attempts have been made from time to time to estimate the Earth's albedo as a function of time by observing the brightness of the dark portion of the Moon's disk, when it is illuminated only as a crescent by the Sun. Due to the difficulty of the method, the results have to be considered very tentative, but they do show evidence for fluctuations of more than 20% in the albedo of the Earth, with a smaller secular increase of about 2.5%, over an 18-month period.[115] If the latter is taken at face value, and if the change was permanent, then the emitting temperature of the Earth would have to fall by around one degree to maintain energy balance, due to this cause alone.

[115] See Earthshine observations of the Earth's reflectance, P.R. Goode et al., *Geophysical Research Letters*, May 2001.

## 10.5  Ocean circulation variations

### 10.5.1  Changes in the thermohaline circulation

In Chapter 5 we saw how the ocean has a major role in redistributing the energy from the Sun from low latitudes, where most of it arrives, to high latitudes that would otherwise be frozen. At first sight, it seems unlikely that this would change, so long as the incoming flux of energy remained the same and was available to drive the general circulation. However, while this may be true of the atmosphere, which shares the task of global redistribution of incoming solar energy with the ocean, we also saw in Chapter 5 that the global overturning of the ocean depends not just on density gradients produced by temperature differences, but also on salinity and surface wind stress.

Most of the sunlight that falls on the ocean is absorbed in the top few metres, so the warmest water is at the top. This is a stable state, but one that increases the importance of salinity gradients for driving the thermohaline circulation. These gradients can be affected by changes in rainfall patterns, or by the injection of relatively fresh water from greenhouse-induced melting of the polar caps. At present, there is a larger imbalance between evaporation and inflow in the Atlantic than the Pacific, which causes the former to be saltier on average. This difference in salinity between the Atlantic and the Pacific is one of the driving forces for the thermohaline circulation (Section 5.6.2). There is reason to believe that this may have been radically altered by salinity changes in the relatively recent past (see Section 10.7), and possibly even that present conditions are tending towards a repeat performance as global warming melts ice from the polar caps.

### 10.5.2  Rapid climate change

While the major ice ages appear to have developed and evolved slowly, over hundreds of thousands of years, the temperature record determined from geological and isotopic data also reveals large changes in the climate that have occurred much faster than this. One of the best-studied examples occurred at around the time the last ice age ended, 10–15,000 yrs ago. This, known as the Younger Dryas[116] event, involved mean annual temperature changes in Greenland of more than 10K only a few decades (Fig. 10.7). Numerous other events of this kind appear to have occurred further back in the past.

The discovery of the Younger Dryas and similar events in the climate record has forced researchers of the paleoclimate to the conclusion that change can be remarkably abrupt. It used to be thought that the large thermal inertia of the ice sheets meant that climate could only change very gradually. This is certainly correct if we rely on radiative processes to melt the ice. However, it is now believed that physical processes in the ice accelerate the melting rate. In particular, the polar ice sheets store a lot of potential energy, and once they start moving, internal friction provides heat that accelerates disintegration. Once separated, ice floes can move to warmer parts of the world. The crust below the ice is generally depressed under the weight of the cap, and will begin to rise when deglaciation starts, accelerating the

[116] The name arose because it was first noted in deposits of pollen in geological strata. These indicate changes from forests to herbaceous plants and back during the ice ages. One of the plants providing key evidence was the *Dryas*, a kind of mountain rose found on glacial tundras, and this is the most recent such event on record, hence 'Younger'. In the early 1980s, evidence for the Younger Dryas was also obtained from $CO_2$ trapped in ice-cores, confirming that it was an event of global significance (since carbon dioxide is well-mixed throughout the global atmosphere).

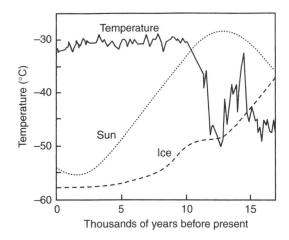

**Fig. 10.7**
The mean temperature in Greenland for the last 17,000 yrs, showing the end of the last ice age and the rapid fluctuations that preceded it. The dotted and dashed lines, respectively, indicate the accompanying trends in the level of insolation on northern midsummer's day (labelled 'Sun'), as moderated by the Milankovich cycles, primarily the precession of the Earth's spin axis (see Fig. 10.1); and the area covered by the N. polar ice cap (labelled 'Ice'). The sharp fluctuation in the temperature that nearly coincides with the maximum summer sunfall is the Younger Dryas event. (Temperature data from: Abrupt increase in Greenland snow accumulation at the end of the younger Dryas event, R.B. Alley, D.A. Meese, C.A. Shuman, A.J. Gow, K.C. Taylor, P.M. Grootes, J.W.C. White, M. Ram, E.D. Waddington, P.A. Mayewski, and G.A. Zielinski, *Nature*, **362**, 527–529, 1993.)

process of disintegration. Glaciers that have their base below sea level are particularly unstable, as are those near rivers and bays that hold relatively warm water. The data shows very rapid warming at the end of Younger Dryas, with very rapid melting leading to sea level rises of as much as 50 m in less than 1,000 yrs, followed by extensive flooding of low-lying areas worldwide.

The cooling phases, which occurred even more rapidly, appear to involve changes in the ocean circulation, since no other plausible mechanism (except possibly enormous volcanic eruptions) could act fast enough, and also because independent evidence from the study of ocean sediments favours this interpretation. The implication is that the thermohaline circulation has more than one quasi-stable mode of circulation, and that the ocean can switch rapidly between them, for example, when a large flux of relatively fresh water from a melting icecap alters the salinity balance. Such 'eigenstates' are found in simple models of the ocean, and something similar may also occur in complex climate models (Section 11.3).

Figure 10.7 includes an indication of how the solar flux in northern summer[117] varied during the period that included the Younger Dryas, showing how it correlates with the temperature record and the geological evidence for the extent of the retreating polar ice cap. The most likely scenario that goes with this is that the increased summer sunfall caused by the Milankovich cycles melted the glaciers that then covered much of North America. Some of the runoff flowed into the Northern Atlantic, and the resulting reduction in salinity was enough to stop the rapid sinking at

[117] Note that here, and in Fig. 10.7, we are talking about the *phase* of the maximum solar input to the Earth that occurs when the planet, in its elliptical orbit, is closest to the Sun, and not the annual mean solar heating. If the Earth is closest to the Sun when it is midsummer in Greenland, then the probability of initiating melting of the icecap is greatest. The fact that this was the case when the Younger Dryas event occurred lends weight to the theory that the event was triggered initially by the shortest of the Milankovich cycles, and then accelerated by changes in the ocean circulation caused by the onset of melting.

high latitudes that draws warm shallow water northwards in the Gulf Stream. In a relatively short time, less than a thousand years, the fresh water supply froze, restoring the Gulf Stream and raising temperatures again.

### 10.5.3 Ocean–atmosphere coupling: The El Niño–Southern Oscillation (ENSO)

The ENSO is a phenomenon driven by coupling between the ocean and the atmosphere in the Pacific, which affects weather and climate over a large part of the globe. It is of particular interest at present because current climate models appear to be on the verge of predicting it successfully. Forecasting anything to do with the weather a few years or decades ahead would be a considerable advance for climate forecasting, promising to provide practical warnings to the affected communities.

The ENSO appears as enhanced sea surface temperatures in the eastern Pacific off the coast of Peru, in a cycle that occurs every 2–7 yrs. This phenomenon was known for centuries to Peruvian fishermen, whose catches were smaller when it happened, because the warm surface waters contain fewer nutrients for the fish than cold water from deep in the ocean. This problem became seriously worse in those years in which the warming started earlier in the year and lasted longer than normal. Because it happened around Christmas, the manifestation became known as El Niño, referring to the Christ Child.

The modern name, ENSO, reflects the fact that we now know that El Niño is part of a global cycle, one that also incorporates another long-observed but not understood phenomenon, the Southern Oscillation. The eponymous oscillation is in the surface pressure difference between the south-eastern tropical Pacific and the Australian–Indonesian region, with corresponding fluctuations in droughts in Australia, floods in east Africa, hurricane/typhoon frequency in the Atlantic and Pacific, and bad harvests in Asia and South America.

The normal conditions in the Pacific Ocean have a warm pool of water in the west, where the warmest water anywhere in the world is found (Fig. 10.8). The prevailing trade winds blow from the east to the west, pushing the warmest water to the western tropical Pacific, leaving colder water to rise off the coast of South America.

In El Niño years, the trade winds relax and can even reverse—this tendency of the winds to switch direction is part of the Southern Oscillation. Then the warm water remains in the east, or is enhanced by warm surface water from the west, and the cold, nutrient-rich water no longer reaches the surface by upwelling (to the distress of the local fishermen). The opposite condition can also occur, in which, instead of reversing, exceptionally strong trade winds blow from east to west, making the water in the eastern tropical Pacific colder than usual. This has become known as La Niña.

The ENSO behaviour has been going on for millennia and its effects can be discerned in the record left in tree rings, ice cores, and coral reefs. They vary considerable in size, intensity and duration; the 1982/83 El Niño was estimated to have cost 2,000 lives and to have done $10 billion of damage worldwide. The 1997/98 event was even stronger, the most intense to be

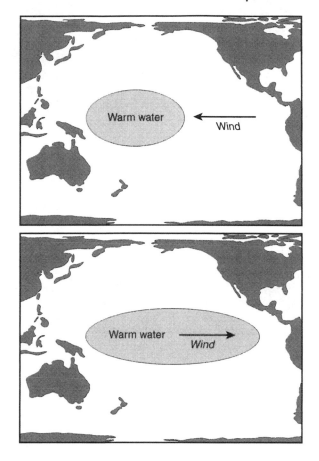

**Fig. 10.8**
The equatorial Pacific under normal conditions (above) and during an El Niño event (below). A slow, planet-wide, irregular oscillation in the prevailing winds tends to pile the warm surface water up towards the west and then, several years later, towards the east.

experienced in the twentieth century, but it was predicted about a year ahead, allowing preparations to be made.

## 10.6   Natural temperature fluctuations

It is a matter of common experience that some years are warmer than others. This is true not only at one observer's location, but (to a lesser extent) in the global average as well. It arises from the climatic version of the chaotic fluctuations that we expect from any complex, dynamic system, from the compound pendulum upwards. In climate research our goal, so far only partially fulfilled, is to be able to separate out the genuinely chaotic fluctuations from those that can be attributed to identifiable cycles (such as ENSO) or to specific forcings (such as the sunspot cycle). We should then like to know the amplitude and the periods of the remaining, unassigned fluctuations, so we could decide whether natural variability is a possible candidate to explain the recent warming trend, instead of (or along with) anthropogenic greenhouse gas emissions.

There are two ways to address this. We can use numerical models to investigate the range of fluctuations we should expect from the climate system, or we can look at the long-term record of the climate itself and attempt to extract the chaotic component.

Since 2003 a project called climateprediction.net has been running experiments to determine the sensitivity of a fairly sophisticated climate model to small perturbations in the initial conditions, all staying well within realistic values for the real climate. The model used is a version of the UK Met Office forecasting model, with the spatial resolution of the atmospheric general circulation model (GCM) coupled to a simplified, single-layer model of the ocean. While not as advanced as the models used in the IPCC analysis, this still has many of the features of the most sophisticated models available and can be expected to give a useful measure of natural climate variability by looking at the spread of results obtained from many runs of essentially the same climate forecast. The current state of this work and its conclusions (plus a chance to participate) are summarized on the project website.[118]

An understanding of actual climate fluctuations and their causes in the distant past is not easy to come by, since before the satellite era there were no truly global measurements, and observations tend to be less reliable as we go back in time. Before about 1800, there were virtually no measurements at all, in the modern sense of using calibrated instruments,[119] and we have to rely on proxies. These are of two kinds: geological, for example air pockets trapped in ice cores, which can be studied with modern instruments and dated by their position in the vertically stratified deposits, and historical, using subjective archival data, such as diaries recording abnormally cold winters, for example.

The latter are actually a surprisingly rich source of information, since climatic fluctuations have always been important in everyday life, leading people such as clergymen and teachers to keep detailed records such things as the first snowfall of winter and the progress of the cycles of vegetation growth and bird migration. In medieval Europe, 'weather diaries' were kept over the active lifetime of many individuals and in a few cases the weather was noted throughout the day, every day. Conditions during important events such as battles or coronations (and, of course, droughts, famines, and floods) were also recorded in literature not intended for scientific use but which is valuable none the less.

The proxy data from natural archives cover the last billion years, approximately, while the period covered by documentary evidence is approximately the last thousand years. By sticking to primary sources—people who were there at the time—and looking for independent corroboration in the form of multiple accounts, researchers have been able to reconstruct rough weather maps for Europe for as long ago as the winter of 1076–1077 (which was bitterly cold). Figure 10.9 shows temperature excursions from the mean for central Europe from 1200 until the present day, reconstructed from this kind of proxy data. If these data are accurate, and if we take them as representative of global mean fluctuations (both big assumptions) then they illustrate why we still cannot be completely sure that the current warming trend of around 0.6°C in 30 years is really of anthropogenic origin and not part of a natural fluctuation with a period of the order of decades or centuries.

[118] www.climateprediction.net/science/index.php.

[119] Among the first scientific sites to make regular meteorological observations was the Radcliffe Observatory in Oxford, starting in 1767 and providing a continuous daily record since 1814.

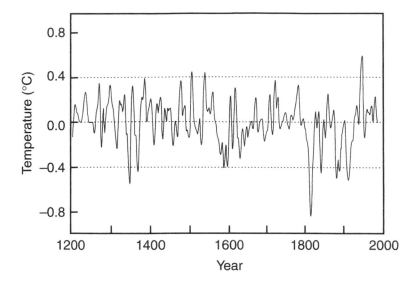

**Fig. 10.9**
Summer temperatures in Central Europe 1200–1990, relative to the long-term mean, from documentary evidence. The corresponding global mean temperatures at these times are not known, but the excursions were probably smaller, if not completely different. (Data from University of Bern, Environmental History Group.)

## 10.7 Summary

From an understanding of past climate change, and the mechanisms involved, we seek a scientific understanding of the present-day climate and its stability. Considering the mechanisms reviewed in this chapter:

 (i) Astronomical (Milankovich) cycles are important in the longer term but do not seem to have any important components on the scale of decades or centuries.
 (ii) Solar variations likewise: the solar constant has not varied by more than 0.1%, definitely in the last 25 yrs, probably not in the last century, and possibly not in the last million years.
(iii) Greenhouse gases have experienced long-term natural and recent anthropogenic changes. The latter match the measured warming trend of the last few decades well if the cooling effect of sulphate aerosols is included.
(iv) Rapid climate change, requiring mere decades to significantly alter land and sea surface temperatures, is possible due to changes in the deep ocean circulation. This has occurred in the relatively recent past (around 10,000 yrs ago) and may be poised to happen again, if triggered by greenhouse warming.
 (v) The coupled ocean–atmosphere system oscillates in ways (like ENSO) that are predictable to some degree and offer the prospect of annual or decadal climate forecasts. The possibility of larger amplitude, slower fluctuations is neither established nor ruled out at present, but none has been found in the temperature record for the last thousand years.
(vi) We have no idea how the Earth's cloud field has varied in the past, and no clear picture of any current trends. We do know that if any trend in cloud amount or properties occurs, it potentially has a large effect on climate.

These effects can all act at once and are coupled to each other. Some (orbital variations) are too long-term to be of concern for present-day climate change; others (perhaps those involving clouds) are quite short, so that they may average out over the course of a season or a year. To understand the net effect of everything we must rely on models, the subject of the next chapter.

## Further reading

*Global Warming: The Complete Briefing*, J.T. Houghton, 3rd edition, Cambridge University Press, Cambridge, 2004.
*Climate Change 2001: The Scientific Basis*, Y. Ding et al. (eds.), Cambridge University Press, Cambridge, 2001.

## Questions

1. Describe the Milankovich cycles, and how they could affect the climate of the Earth. What is the predicted amplitude of all three cycles considered together?
2. Use a simple model to calculate the effect of a change of this magnitude in the solar energy intercepted by the Earth on the mean surface temperature in a simple model. What factor makes this of less concern to humans than other changes thought currently to be affecting the climate?
3. List the four main factors that are known to affect the climate at the Earth's surface, and discuss briefly the past, present, and future stability of each, with particular emphasis on the possible timescales for producing significant climate change in each case.
4. Using a simple model, calculate the effect of a 1% change in (i) Earth's albedo, (ii) the output from the Sun, on the surface temperature. How likely are these to be equal and opposite?

# Climate models and predictions

**11**

## 11.1 Introduction

Studies of climate and climate change, like those of virtually all problems involving complex coupled systems and a multitude of different processes, rely on models.[120] Simple models provide useful insight into the physics involved, while complex models are necessary to obtain the greatest possible accuracy and to attempt to incorporate all of the feedback process.

The simplest climate models are zero or one-dimensional energy balance models, like those introduced in Chapter 7. These have low computational costs and explicit formulations, so that the links between individual parameters or processes and the climate can be seen and their dependence evaluated. Complex climate models are more realistic but computational expensive, and contain so many interdependent variables and relationships that it is usually not possible to follow cause and effect directly. They are based on general circulation models (Section 1.10) for the atmosphere and the ocean, coupled together, and have subroutines that treat ice formation, the carbon cycle, chemical processes, land and vegetation processes, in particular those affecting atmospheric composition and surface reflectivity.

In seeking to use models to predict the future evolution of the climate system with some reliability, close attention obviously has to be paid to the accumulation of errors. These are of (at least) three kinds:

(1) those due to limitations in the model itself, for example, in spatial or temporal resolution, numerical methods, or the data used for initialization;
(2) those due to approximate, wrong, or missing physics in the model; and finally
(3) those due to the macroscopic equivalent of the uncertainty principle: the inherently chaotic nature of the climate system.

The last of these may be the most formidable, since there is still no consensus about which aspects of climate are predictable on any particular space or time scale.

Once physical models of the climate have been set up, it becomes possible to explore the effect of changing the external and internal forcing. These can include the seasonal cycles changes due to small, periodic changes in the Earth's orbit (the Milankovich cycles, see Section 10.2), which are thought to be responsible for the ice ages, and changes in the minor constituent composition of the atmosphere due to pollution of human origin, which

[120] Models are simplified representations of complex phenomena. They reduce the system under investigation to a set of one or more equations that can be solved either analytically or numerically.

alter the greenhouse warming. But before using a model to predict the response to future changes of these kinds and others, we must of course make sure that it correctly reproduces at least the most important features of the present day climate, and that it tracks changes that have taken place in the past.

The details of the formulation of complex, advanced models is beyond our present scope, but we will consider what physics is involved, what their most recent results show, and what they predict for the future. Again, we make use of the work of the Intergovernmental Panel on Climate Change (IPCC) of the World Meteorological Organization to obtain a convenient and reliable synthesis of the results of a large number of research groups. A great deal of work has been done by the IPCC to coordinate the work of the scientists with the best models, all over the world, so that their results can be averaged to give a 'most probable' prediction, and the scatter of their results and estimated error bars examined to put a figure on the uncertainties in this prediction.

As already noted, one of the first tasks of any such team intent on producing useful forecasts is to show that the tools they are using can successfully account for the known climate record covering as far back into the past as possible, a procedure sometimes known as 'hindcasting.' Figure 11.1 summarizes the consensus on global mean temperature changes in the recent past.

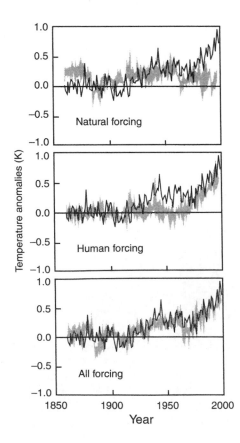

**Fig. 11.1**

Observed annual global mean surface temperatures (black line) compared to simulations by an ensemble of climate models (grey shading) by the IPCC. Three sets of models are compared to the same observational data, assuming (a) no perturbation of the atmospheric composition by human activities, but including natural events like volcanic eruptions; (b) human influences only, and (c) both. The last of these provides the best fit to the data and also shows that the change in the mean surface temperature has, in the last two decades, started to exceed the short-term natural fluctuations (the 'noise' on both the measured and model temperature plots).

Note three things:

(1) The chaotic fluctuations in the models and in the data show similar amplitudes;
(2) The observed global temperature has, since about 1980, risen beyond the range of natural variability, at least for fluctuations with periods of less than 150 yrs;
(3) Anthropogenic forcing explains most of the increase, and, when known natural forcings (e.g. solar variations and volcanic eruptions) are included as well, an excellent fit between theory and observation can be obtained. We see that the anthropogenic is larger than the natural contribution in the last 20 yrs.

## 11.2 Current models and predictions

The basic building blocks of the coupled atmospheric and oceanic general circulation models are the primitive equations,[121] which express momentum balance, hydrostatic balance, conservation of mass, and conservation of energy. To these must be added schemes for radiative transfer, cloud formation, surface interactions, and all the processes described in the preceding chapters. These frequently involve the simplification, for example, by linearization, of complex processes, and constantly demand improvement. The resulting models all give a heavily damped, essentially linear response to changing boundary conditions, which is fairly unrealistic especially for long-term forecasts, and explains why models cannot at present reproduce the observed interannual variability.

We have seen (e.g. Figs 10.6 and 11.1) that model calculations have been quite successful at accounting for recent greenhouse warming of the order of 1K. This gives us some confidence that the models contain enough of the right physics to allow us to make cautious predictions about the future behaviour of the greenhouse effect. Figure 11.2 shows the different phenomena thought to be currently forcing climate change, both warming and cooling, with IPCCs best estimate of their relative importance, and the associated uncertainty.

[121] There are three primitive equations: the (vector) momentum equation, the continuity equation, and the thermodynamic equation. The first of these is basically Newton's second law, with terms for gravity, pressure gradients, Coriolis force, and friction; in its simplest form it just represents hydrostatic balance. The thermodynamic equation is a statement of the first law of thermodynamics (conservation of energy) with terms for changes in internal energy (of an air parcel) due to adiabatic heating, sensible heat fluxes (including radiation), and latent heat exchange.

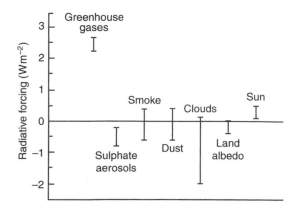

**Fig. 11.2**
The IPCCs estimate, with uncertainties, of the relative importance of greenhouse gases, clouds, and aerosols, changes in land use (e.g. deforestation) and changes in the Sun, on current trends in global warming. Each is expressed as a radiative forcing in $W\,m^{-2}$, equivalent to a change in absorbed solar flux of this amount.

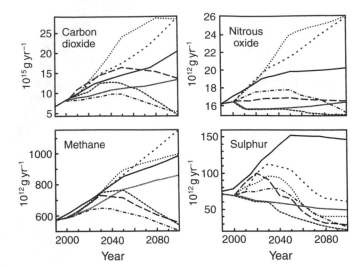

**Fig. 11.3**

Anthropogenic emissions for three key greenhouse gases and sulphur (the progenitor of sulphate aerosols) projected into the future, according to seven possible 'scenarios' adopted by the IPCC.

Even if the models were a perfect representation of the real climate system, however (which of course they are not), a further uncertainty is how the accumulation of greenhouse gases in the atmosphere will evolve over the time scale of interest. The future trends in each of these can be expressed as a function of time: for example, in Fig. 11.3 we show a set of assumptions or 'scenarios', which the IPCC adopted for greenhouse gas emissions up to the end of the present century. They vary substantially, depending on the degree of optimism adopted for the development and application of 'clean' technologies, legislation to control emissions and so forth.

Figure 11.4 shows a summary of the results of using these scenarios to obtain a range of model predictions for the mean surface temperature, over the same 100-yr span. The variations between the scenarios are represented by the heavily shaded area; the lightly shaded area is the range when the differences between model predictions for the same scenario are also folded in. The best estimate, based on this figure, of the global warming in store for us is about 3K by the year 2,100, less than a century away. Allowing for the uncertainties, mainly in the extent to which negative feedbacks will limit the upward forcing due to $CO_2$ and other greenhouse gases, which may themselves be wrongly estimated, it could be as low as 1K or as high as nearly 6K, but most probably between 2 and 4K.

These are fairly serious numbers, which have as one of their more important consequences a projected rise in global sea level. A very simple calculation illustrates this. The mean depth of the oceans is about 5 km and the coefficient of thermal expansion of water is approximately $2 \times 10^{-4} \, K^{-1}$ at room temperature. If this coefficient is simply assumed to apply to all of the water in the ocean, and if it is further assumed that any temperature rise at the surface is rapidly communicated to the ocean depths (neither of which is a very accurate assumption, in fact) then it follows that each degree of global warming will result in a sea level rise of 1 m. Despite the grossly simplified calculation, this illustrates the scale of the problem we can expect if the expected warming of 3° or 4° materializes in the present century.

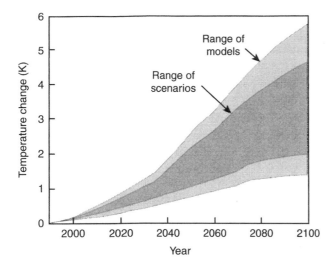

**Fig. 11.4**
A summary of the IPCCs predictions of global warming in the rest of the present century. The 'range of scenarios' covers estimates of the different possible profiles for anthropogenic emissions in the twenty-first century, while the lighter-shaded area represents the range of predictions due to differences between the models themselves.

Added to thermal expansion is the problem of the melting of the polar icecaps. The physics is simple—ice that is floating in the sea, like the Arctic polar cap, makes no difference to the sea level if it melts. Ice that is presently on land, of which the most important examples are the Antarctic icecap and the sheets covering Greenland, does increase the sea level if it melts.

Antarctica has about 90% of the world's ice in a layer approximately 2 km thick, covering an area of about 13.6 million km$^2$. If all of this melted, sea levels around the world would rise by about 60 m. However, the temperature would first have to rise to the freezing point; at a current average temperature of $-37°$C there is no prospect of that in the foreseeable future. The effect of smaller amounts of short-term warming is likely to be an accelerated rate of shedding of large floes of ice into the ocean from the edges of the cap. Once liberated in this way, the floes tend to drift to warmer waters, where they melt fairly rapidly.

The same calculation for Greenland shows that the ice there would raise sea level by 7 m if it melted, and since the current mean temperature is significantly warmer, this is a real prospect. It has been estimated that after a rise in the temperature there of just 3°C, within the IPCC predictions for the present century, the entire ice sheet would be doomed. The domed shape of the Greenland icecap tends to accelerate melting by releasing potential energy and by allowing ice to slide to lower, warmer levels and into the sea.

As with the atmospheric greenhouse, feedback loops may ameliorate the problem: we can imagine, for example, enhanced moisture from the warmer tropics increasing the precipitation of water as snow in the polar regions, offsetting the loss rate in, or even increasing, the mass of the polar icecaps. On the other hand, loss of ice lowers the albedo and accelerates melting. Trying to take all factors into account, the IPCC has estimated that a sea level rise somewhere in the range from 15 to 95 cm is to be expected in the next 100 yrs. They note that the global average temperature increase of $0.6 \pm 0.2$ K in the twentieth century was accompanied by a global average sea level rise of 10–20 cm, measured by tide gauges during the same period. Satellite and other observations show a decrease in the snow and ice cover worldwide. On the other hand, some important climate indicators appear not to have changed, at least in recent decades when good measurements have been available, most notably the mean temperatures in some parts of the southern hemispheric oceans and Antarctica.

In summary, the experts and their models are predicting an *increase* in mean surface temperature in the next 100 years that is comparable to the net *decrease* that the world experienced during the last ice age. There are many uncertainties, of course, but it cannot be assumed without reason that these uncertainties make the problem *less* serious. One of the missing capabilities in current models is the ability to represent multiple stable states and possible transitions between them. If these occur, they are likely to be more serious than gradual changes, even if their amplitudes are no larger, because of the lack of time to prepare for them and adapt to the consequences.

## 11.3  Multiple climate equilibria and sudden climate change

The first generation of advanced climate models emphasized *gradual* change, for example, the steady increase in global mean temperature that results from the build-up of greenhouse gases, ameliorated by a simultaneous build-up in sulphate and other kinds of aerosol, and possible changes in the amount and type of cloud cover. A more recent goal is the modelling and prediction of *abrupt* climate change, in which large changes may occur in only a few decades. The challenge is to understand what the stable states of the climate are, how stable the current state is, what transitions are possible, and how and when they are likely to occur.

This new focus has come about because (a) definite evidence has emerged that rapid climate change, such as the Younger Dryas event (Section 10.5.2), actually occurred in the relatively recent past (around 10,000 yrs ago), and (b) modelling of complex systems in general suggests that the climate system may have multiple equilibria or 'eigenstates', which are stable, separated by unstable or metastable versions that prevail, if at all, relatively briefly. The implication is that climate change may involve gradual evolution for a while, but, once the limits of a stable state are reached, it will then make a rapid transition to the next stable state, which may be quite different. Examples of this kind of climate 'flip' might be from a mainly ice-free planet to one that is heavily glaciated, or a gradual warming of a couple of degrees over a century, say, followed by a much faster increase to a Venus-like state (the 'runaway greenhouse' scenario).

Two simple models, one radiative and the other dynamical, both of which represent some aspects of the real climate system, will be considered to show how multiple equilibria can occur, even in uncomplicated systems. They arise basically because of nonlinearities in the dependences between variables, analogous to the behaviour of quadratic and higher-order equations, which have multiple solutions. This will become clear from the following examples.

### 11.3.1  A simple radiative model

Recalling the simple energy balance model of Section 7.4, but now considering the situation where the mean surface temperature of the Earth is changing at a rate $\frac{dT}{dt}$, we can write:

$$C\frac{dT}{dt} = \frac{S}{4}(1 - A(T)) - e\sigma T^4.$$

This expression equates the energy gain or loss corresponding to the change in temperature $T$, where $C$ is a representative heat capacity, to the difference between solar energy received, $S$, and blackbody radiation lost to space according the Stefan-Boltzmann law as before. The mean surface albedo $A$ is a function of $T$ since the amount of ice on the surface of the Earth will increase or decrease as the temperature changes. The exact form of the function is not very important for present purposes; for simplicity, let us assume that this is a linear relationship until the planet is completely ice covered or completely ice free, when it becomes constant (Fig. 11.5).

At equilibrium $dT/dt = 0$, so

$$S(1 - A(T)) = e\sigma T^4,$$

which can have three solutions for $T$ (Fig. 11.5) corresponding to an ice-covered world, one with no ice and something in-between.

Upon examination, we find that the intermediate solution—the one that most resembles the present-day Earth—is unstable. If $T$ is nudged from this point towards a higher value, it is in the regime where $\sigma T^4 < S(1 - A)$, so $dT/dt > 0$ and $X$ tends to increase, which means a decrease in $A$. This further increases $T$ and amplifies the perturbation until the system reaches a new equilibrium at the high temperature, ice-free solution. If the intermediate solution is perturbed towards *lower* temperatures, the opposite happens to reduce the temperature until it reaches the cold, ice-covered solution.

This result has no great significance for the real Earth, of course, since only one of many feedbacks is represented in the model, and that too quite

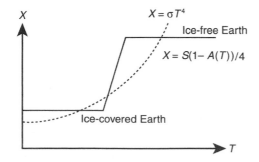

**Fig. 11.5**

Multiple climate equilibria in a simple radiative model. The curve represents the energy leaving the planet and the series of straight lines the energy absorbed, as a function of temperature $T$. The latter depends only on the albedo, so is independent of $T$ when the Earth is fully ice-covered or free of ice.

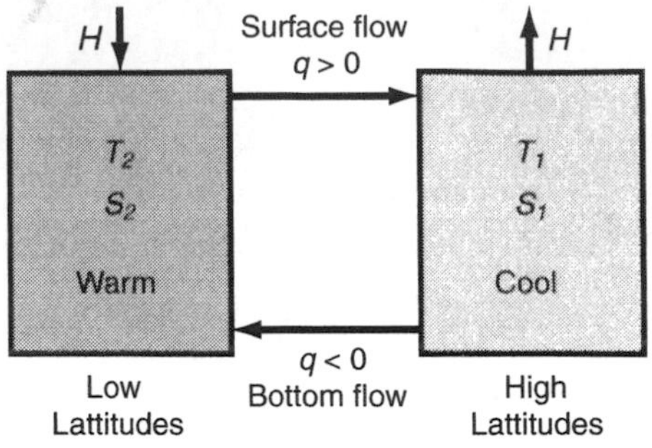

**Fig. 11.6**
A two-box model of the ocean circulation. The temperatures $T_1$ and $T_2$ are fixed so the principal variable is the salinity, $S$.

crudely. It does show, however, how multiple equilibria, some stable and some unstable, can arise. In a complex system like the real climate, we are likely to encounter quasi-stable equilibria as well. These, like a marble on a shelf, are stable against small or short-lived perturbations, but can respond catastrophically to larger or more sustained forcing. Our present-day climate could be one of these.

### 11.3.2　Box models of the ocean

In Section 10.5 we discussed how sudden climate change might be caused by transitions between quasi-stable states of the thermohaline circulation in the ocean. We now look to see if this can happen in a simple model with some realistic features. The two-box model first studied by Stommel[122] is illustrated in Fig. 11.6.

The two reservoirs of well-mixed water connected by pipes represent the polar and equatorial regions of the ocean at temperatures $T_1$ and $T_2$ respectively. For simplicity, these temperatures are fixed in the model, equivalent to assuming that the oceans are in equilibrium with the solar flux and the surrounding land mass. The principal variable is the salinity of the water, $S$, which is affected by a flux $H$ of through the atmosphere. $H$ is not an actual flux of salt of course, but the flux of salt-free moisture composed of the net effect of evaporation and precipitation. This is equivalent to a 'virtual' flux of salt in the opposite direction, which is positive when transfer is from box 2 to box 1. This should be the normal state of affairs since $T_1 < T_2$, which tends to make low latitude water saltier than that at high latitudes. The flow of water $q$ is the same in both pipes and is considered positive when polewards in the upper pipe, again corresponding to the normal behaviour in the real ocean.

If the flow between the boxes is directly proportional to the density difference and hence proportional to the difference in the two values of $S$, conservation of salt is expressed by

$$\frac{\mathrm{d}S_1}{\mathrm{d}t} = -H + |q|(S_2 - S_1)$$

and

$$\frac{\mathrm{d}S_2}{\mathrm{d}t} = H + |q|(S_1 - S_2),$$

[122] Henry (Hank) Stommel (1920–1979) was Professor of Physical Oceanography at MIT, and also held positions at Harvard University and Woods Hole Oceanographic Institute. In 1961 he published 'Thermohaline convection with two stable regimes of flow', *Tellus* **13**, 224–230.

where $V$ is the volume of a box, and the flow rate between boxes $q$ is represented by its modulus or numerical value, since the equation is valid whether $+q$ or $-q$ is inserted. Representing the resistance to the flow by a linear frictional term with coefficient $k$ we can write

$$q = \frac{k}{\rho_0}(\rho_2 - \rho_1),$$

where $\rho_0$ is a reference density.

We also require an equation of state for the water, which can be chosen to be

$$\rho = \rho_0(1 - \alpha T + \beta S),$$

where $\alpha$ is the thermal expansion coefficient and $\beta$ the haline 'contraction' coefficient (so called because an increase in $S$ acts on the density in the opposite sense to an increase in $T$, hence the different signs since both $\alpha$ and $\beta$ are positive constants).

Now let

$$\Delta T = T_2 - T_1$$
$$\Delta S = S_2 - S_1$$
$$\Delta \rho = \rho_1 - \rho_2$$

(note the different sense of the definition of $\Delta\rho$). Now we have

$$q = \frac{k}{\rho_0}\Delta\rho = k[\alpha(\Delta T) - \beta(\Delta S)]$$

from which we see that temperature and salinity gradients tend to drive the circulation in opposite directions. We also have

$$\frac{dS_2}{dt} - \frac{dS_1}{dt} = 2H - 2|q|(\Delta S)$$
$$= 2H - 2K|\alpha(\Delta T) - \beta(\Delta S)|\Delta S$$
$$= 0$$

for a steady-state.

There are three steady state solutions to this equation, two of which are stable.[123] If we look first for solutions with $q > 0$ and $\alpha(\Delta T) > \beta(\Delta S)$, similar to the real situation, then

$$2H - 2k(\alpha\Delta T - \beta\Delta S)\Delta S = 0$$

or

$$[\beta\Delta S]^2 - \beta\Delta S \alpha\Delta T + \beta H/K = 0,$$

which has roots

$$\beta\Delta S = \alpha\Delta T\left(\frac{1}{2} \pm \sqrt{\frac{1}{4} - \frac{\beta H}{k(\alpha\Delta T)^2}}\right)$$

[123] Again, 'stable' means that the model ocean will tend to return to the solution if perturbed away from it. For unstable solutions, the 'ocean' would continue to move away until it found another solution that was stable. Some models, usually more complex ones, can have 'metastable' solutions that are stable against small perturbations and unstable if the perturbation is large, or sustained, enough.

corresponding to 2 'thermally direct' solutions (i.e. solutions with motion from the region of greater to lesser heating) that exist if

$$4\beta H < k(\alpha \Delta T)^2.$$

Suppose we simplify this expression by defining two new variables:

$$\delta = \frac{\beta \Delta S}{\alpha \Delta T},$$

essentially a non-dimensional salinity difference, since $\Delta T$ is a constant, and

$$E = \frac{\beta H}{k(\alpha \Delta T)^2},$$

which is a dimensionless salinity flux. Substituting, we get for the two solutions

$$\delta_1 = \frac{1}{2}(1 - \sqrt{1 - 4E})$$

and

$$\delta_2 = \frac{1}{2}(1 + \sqrt{1 - 4E})$$

that exist if $4E < 1$.

We can also look for 'thermally indirect' solutions, that is, those for which $q < 0$ and $\alpha(\Delta T) < \beta(\Delta S)$]. Then we have

$$H + k(\alpha \Delta T - \beta \Delta S)\Delta S = 0$$

or

$$[\beta \Delta S]^2 - \beta \Delta S \alpha \Delta T - \beta H/k = 0,$$

which has only one root with positive salinity:

$$\beta \Delta S = \alpha \Delta T \left( \frac{1}{2} + \sqrt{\frac{1}{4} + \frac{\beta H}{k(\alpha \Delta T)^2}} \right)$$

or

$$\delta_3 = \frac{1}{2}(1 + \sqrt{1 + 4E}).$$

corresponding to a 'salinity driven' solution for the circulation, that exists for all $E$.

All three solutions are shown in Fig. 11.7, along with the trends that result from small perturbations to the value of $\delta$. Solutions with $\delta > 1$ and $\delta < 0.5$ correspond to the situation where salinity and temperature, respectively, dominate the flow. The temperature-dominated circulation has the flow in

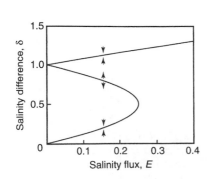

**Fig. 11.7**
Dimensionless salinity difference $\delta = \beta \Delta S / \alpha \Delta T$ plotted against dimensionless surface salinity flux $E = \beta H / k(\alpha \Delta T)^2$ showing equilibrium solutions (lines) and tendencies (arrows). The latter indicate that the middle branch of the curve consists of solutions that are unstable.

the sense from pole to equator in the bottom pipe, and corresponds to the configuration in the present climate where we find upwelling in the low latitudes and the surface flow travelling towards the poles.

The salinity-dominated circulation, solution $\delta_3$, has the flow reversed, with salty water sinking in the warm box, and flowing at depth towards the poles. Cold polar surface water then flows equatorwards, although with smaller values of $q$. Physically, what this is saying is that the 'reversed' circulation is more sluggish than that in the 'normal' situation, since the evaporative processes responsible for building up salinity differences are slow, and rapid flow would restore the dominance of temperature.

One of the temperature-dominated solutions, that at intermediate values of $\delta$, is unstable against small perturbations in either the $+\delta$ or $-\delta$ direction, as the arrows in the figure indicate. If we examine the graph, we see that increasing $\delta$ requires also increasing $E$ on this branch. Looking at the formula, $\delta_1 = \frac{1}{2}(1 - \sqrt{1 - 4E})$, we see that increasing $E$ further increases $\delta_1$, so the loop represents positive feedback that drives the model climate up to the next stable solution, which is the salinity-driven version. Similarly, reducing $\delta_1$ increases $E$ and drives the climate down to the stable temperature-driven solution.

Again, we cannot equate this unstable behaviour of the solution that most resembles the present-day Earth directly to the real climate, because the model is too simple. However, the possibility exists that metastable states in the climate may transit rapidly to other states. The Stommel model does illustrate one process by which that might occur: greenhouse warming of the polar caps, releasing enough fresh water to stop sinking of the polar waters by reducing their density. Despite their higher temperature, the equatorial waters are then the more dense, and sinking there forces rising at high latitudes—a thermally indirect circulation. In reality, the temperatures in the two bodies of water would not remain the same in this situation, and it is easy to envisage extensive glaciation as a result.

### 11.3.3  Complex models

An obvious question to ask next is whether fully fledged climate models show this sort of behaviour or not. This an issue that is right at the cutting edge of current research, so a conclusive answer is not yet available, but we can look at the recent state-of-play.

The most complex climate models include detailed general circulation models for both the ocean and the atmosphere, with refinements such as the topography of the individual ocean basins represented at least approximately. The two models are coupled together via the action of the winds on the ocean surface (which depends, among other things, on the variable wave height), the transfer of heat and moisture between the ocean and the atmosphere, and a host of other processes occurring at the air–sea interface including production of aerosols by sea-spray, and chemical transport between the air and water. Detailed investigation of such processes is the subject of much current work, as is the improved coding of the GCMs themselves, on ever more powerful processors. The principal

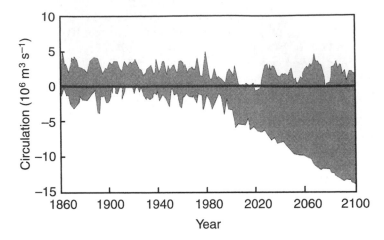

**Fig. 11.8**

The range of predicted values for the change in the flow rate in the thermohaline circulation in the Atlantic Ocean with time, as predicted by the nine different advanced climate models studied by the IPCC. The shaded area shows the spread between models, associated with the modelled natural variability from 1860 to 1960, and then diverging significantly as global warming takes hold in the late twentieth century and beyond.

limitations of the best current models are in three areas: (i) limited knowledge of the physics of key processes, (ii) limited ability to resolve even known processes adequately in space and time, (iii) the inherently chaotic nature of the climate system. Modellers must always introduce approximations, since even a complete set of equations must be reduced to finite differences or coefficients to obtain a numerical solution, and some key processes will remain unresolved, operating at subgridscale levels and requiring a simplified representation or 'parameterization'.

Figure 11.8 shows the past and predicted future flow in the thermohaline circulation in the Atlantic Ocean, equivalent to the flow rate $q$ in the 2-box model, according to an ensemble of nine advanced models. On average, these are predicting that the thermohaline flow rate will decrease substantially during the course of the present century, but they all stop short of a sudden 'flip' in the circulation, and do not reverse. Some of the models show a decline of more than 50% in the N–S flow, which at present is about 20 Sverdrups;[124] others show the circulation continuing virtually unchanged. None of them shows a sharp drop, comparable to the Younger Dryas event, but this is probably due to the inability of the models to represent fast transitions, rather than any reassurance that such a change could not occur. We can see in the historical record (e.g. Figs 10.3 and 10.7) that the climate has changed abruptly in the past. The reason similar events do not appear in the future projections may be that even the best models include simplifications (linear approximations to some non-linear equations, for example) and damping and smoothing that prevent multiple equilibria from emerging as decisively as they do in the simple models of the previous sections. Model improvements in the future will address this, and in any case, even a gradual reduction in the thermohaline flow of the magnitude indicated by an average of all of the current predictions would be serious if it occurred.

[124] Physical Oceanographers call one million cubic metres per second, the unit by which they measure flow in the ocean, one *Sverdrup* (cf. Section 5.6.3).

## 11.4 Problems of detection, attribution, and prediction

Like all complex systems, the climate exhibits internal variations that are neither due to external forcing nor to permanent internal changes. We see some of this in the weather we experience every day; some longer-term changes lasting years or decades are also due to noise in the climate, the random 'sloshing' of the coupled atmosphere–ocean system. Climate models mimic this behaviour, and in principle can allow us to decide when forcing, natural or human, is tending to produce secular change, and whether changes in solar input, eruption of volcanoes, or increased concentrations of greenhouse gases and aerosols, are responsible. With advanced analysis techniques, more extensive and more accurate observations of forcings and responses, and a more complete representation of the relevant physics in the models, it is now becoming possible to demonstrate whether an observed variation in climate and its links to a specific, postulated cause are statistically significant.

The present state of the art in making such calculations show that climate forcing by changes in solar irradiance and volcanism have caused measurable fluctuations in global and hemispheric mean temperatures, but that these and other natural forcings produce too little temperature change to explain the observed rise in the twentieth century. (In fact, the trend in net solar plus volcanic forcing has been negative in recent decades, and, considered in isolation, would tend to produce cooling.) Similarly, the observed decline in the area of sea ice in the Arctic, amounting to nearly 20% in the last few decades, fits global warming model predictions but would be extremely improbable if attributed solely to natural fluctuations (Fig. 11.9).

Model fitting of this kind is a powerful technique for helping to decide whether a change can be attributed to a particular cause. It can be extended

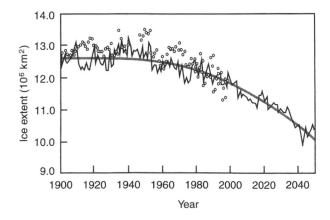

**Fig. 11.9**

Observed (points) and modelled (lines, one with 'natural' fluctuations and the other smoothed) variations of annual average sea ice extent in the northern hemisphere (in millions of sq. km) versus time over the last century, and predicted until 2050. The fit between the two, and the fact that the trend is beginning to exceed the natural fluctuations, represented by the scatter in the data, strongly suggests a long-term trend, which in the model at least is due to greenhouse warming. (NOAA/Geophysical Fluid Dynamics Laboratory.)

to consider more than one variable at a time, which in principle should more strongly discriminate between different contributing and competing effects, and deliver more robust conclusions. If a climate model can forecast not only the global mean change in temperature, or some other climate variable, but also the geographical, seasonal, and vertical patterns, then a successful correlation with the data is much more convincing. If not only the mean temperature but also the global three-dimensional pattern correspondences are found to have been increasing, the probability that these correspondences could arise by chance, rather than forcing, is very low.

A simple example of this kind of pattern analysis that has already proved effective is the simultaneous observation of atmospheric temperature trends in the stratosphere and the troposphere. Model calculations and observations show that increases in greenhouse gases lead not only to a warming of the troposphere and the surface, but also to a simultaneous *cooling* of the stratosphere. Finding both of these changes at once, with about the expected magnitudes and in the right (i.e. opposite) directions, is obviously a more convincing detection of global change than fitting either one individually.

Figure 11.10 illustrates how this works conceptually. The $x$ and $y$ axes represent changes in the mean temperature of the troposphere and the stratosphere, respectively, over some sufficiently long period, say $\geq 20$ yrs. Suppose satellite soundings are used to obtain the mean temperature of the whole troposphere, averaged vertically and with latitude and longitude, over all days in 1980, and the same thing for the stratosphere. Then we do it again for a later year, say 2000. The difference between the two mean tropospheric temperatures provides the value on the $x$-axis, and the difference for the stratosphere the $y$-axis value, so we can plot a data point on the graph. If it falls at the origin, this particular measure of global change is zero.

The two shaded areas in the plot represent the natural variability of the atmosphere and model predictions respectively. These can be obtained

**Fig. 11.10**

A change in two or more climate-related variables (here the mean stratospheric and mean tropospheric temperatures) that is (a) in the right direction for both, and (b) greater than the natural fluctuations observed in each variable, and (c) comparable in size and direction to that predicted by a comprehensive climate mode, can be attributed with a quantifiable level of confidence to the anthropogenic forcing applied in the model to produce this change (here an increase in atmospheric $CO_2$ abundance).

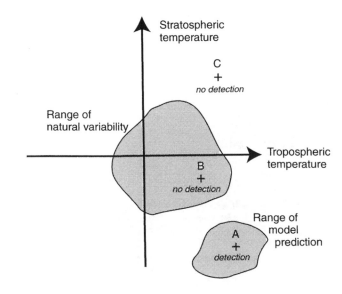

(again this is just an example, as there are many ways of doing it) by forming *daily* averages of the tropospheric and stratospheric temperatures in 1980 and in 2000, taking 366 twenty-yr differences, and plotting the mean (shown as a cross, +) and the deviation from the mean as an envelope that contains all of the points (this is the blob labelled 'range of natural variability'). We then repeat the exercise with the model data, calculated with the observed amounts of greenhouse gas for each year, and this time plot the envelope of the predictions without removing the mean (blob labelled 'range of model predictions').[125]

For a positive detection of global change, the measured mean change should fall *outside* the 'natural variability' blob, and *inside* the 'model predictions' blob, like point A in the diagram. If we obtain point B instead, we cannot claim a detection, because although change has been observed, it is smaller than the natural fluctuations and therefore we can only suspect a changing greenhouse effect. Had we obtained point C from the measurements, corresponding to large increases in *both* temperatures, we would conclude that global warming was occurring but not by the mechanism that we presently understand, that is, not due to increasing greenhouse gases. We would begin at that point to suspect something like increased solar output instead, that might warm both the troposphere and the stratosphere together.

The position in global change detection, based on this sort of analysis (with much more sophistication, of course, including comparing many different time periods), was that points similar to B were found until quite recently. This suggested greenhouse warming but did not prove it (even in the limited sense of this one kind of statistical test). The 'model predictions' blob also overlapped the 'natural fluctuations' blob, indicating that a definite result might not be possible, in other words that the warming expected from models was within the range of the observed natural fluctuations. In the last few years, however, the warming trend has grown so that the blobs have moved apart and points like A are now being found.

The point is that statistical analyses based on more than one variable (two, in this case) provide a stronger indication of greenhouse-induced warming than surface temperature data alone. From this relatively simple example, we can go on to consider the state in the art in matching more detailed global patterns, as illustrated in Fig. 11.11. The principle is the same, but now the atmosphere is divided into many regions, and advanced pattern recognition techniques are used to compare the model predictions to observations.

The example in Fig. 11.11 shows that a model that calculates the temperature change expected from changes in manmade greenhouse gases alone reproduces the difference in response either side of the tropopause reasonably well, but with many detailed discrepancies. Including the effects of the observed amounts of anthropogenically produced sulphate aerosols and stratospheric ozone enhances the contrast between the troposphere and the stratosphere, and (although it may not be very obvious to the eye) improves the fit with the data, providing a better match than one that includes greenhouse gases alone. Although the fit is still far from perfect, analyzed statistically it does correspond to quite high confidence in model predictions of greenhouse warming, well outside of the natural variations observed during the same period. The details of the imperfections in the fit contain

[125] Realistically, we would probably draw the envelope around 99%, or perhaps 95%, of the points, to prevent 'rogue' data from distorting the envelope too much, but this requires care. In any case, it is interesting in an experiment like this to compare the two shaded areas for size and shape, since one represents the observed, and the other the modelled, 'natural variability' of the climate.

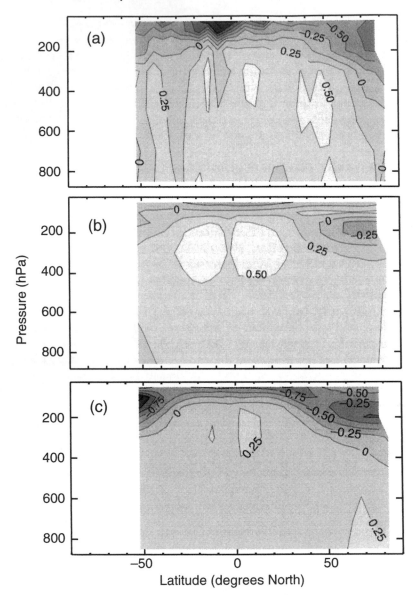

**Fig. 11.11**
Simulated and observed zonal mean temperature change as a function of latitude and height. The contour interval is 0.25°C. All signals are the difference between the 1986 to 1995 decadal mean and the 20 yr 1961 to 1980 mean. The three frames show: (a) the measured data; (b) model simulations, in which changes in the greenhouse gas concentrations between the two periods were taken into account; and (c) the same model simulations but with observed changes in stratospheric ozone in sulphate aerosols also included. M.R. Allen and S.F.B. Tett, Checking for model consistency in optimal fingerprinting, *Climate Dynamics* **15**, 419–434, 1999.

information useful for improving the models, the measurements, and the comparison techniques.

## Further reading

As Chapter 10, plus:
*Climate System Modelling*, K.E. Trenberth, Cambridge University Press, Cambridge, 1992.
*Can we believe predictions of climate change?* J.F.B. Mitchell, *Quarterly Journal of the Royal Meteorological Society*, **130**, 2341, 2004.
*Climate Prediction* Website: www.climateprediction.net/index.php

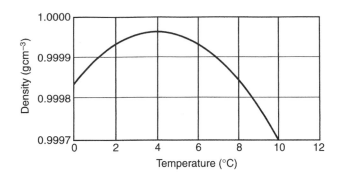

**Fig. 11.12**
The density of water as a function of temperature for constant salinity.

## Questions

1. Describe the three main categories of error in models used to forecast future changes in the climate. How might these be (a) reduced, (b) quantified?
2. Using the data in Fig. 11.12, and making suitable assumptions about the other properties of the ocean, estimate the mean global sea level rise in the next 50 yrs if the mean temperature of the atmosphere near the surface increases by 5K in that time. If the mean atmospheric temperature remained constant for a further 500 yrs, how much more would the mean global sea level rise? Would it continue to increase after that, and, if so, why, and by how much? Discuss briefly how realistic this approximation is, and what factors in the real ocean–atmosphere system might alter the result.
3. Show that a simple radiative equilibrium model of the Earth, in which the albedo of the planet is temperature-dependent, can have more than one stable state.

   In such a model, with albedo $A(T)$ where $T$ is the temperature of the Earth, the coldest stable state occurs for $T = 200$ K and the warmest for $T = 275$ K. Calculate the value of the albedo in each case.
4. Describe a simple 2-box model of the ocean circulation and show how this can also have multiple solutions. For such a model with salinity difference $\Delta S$ and temperature difference $\Delta T$, show that solutions of the form

$$\delta = \frac{1}{2}(1 \pm \sqrt{1 \pm 4E})$$

   are predicted, where $E$ is a dimensionless representation of the salinity flux. Discuss the possible significance of these four solutions in the real climate system.
5. Visit the Climateprediction.net website and then summarize: (a) what this project aims to achieve, (b) the methods being used, and (c) current results.

# 12 Climate on other planets

## 12.1 Introduction

Climate systems analogous to that which concerns us on the Earth are found on the other planets of the Solar System. The processes are the same or similar, and the physical laws identical; it is the different boundary conditions, especially distance from the Sun, that give rise to different histories and present-day solutions to the equations governing climate.

Not surprisingly, the regimes on the nearby, and similarly sized planets Mars and Venus, bear the closest resemblance to the Earth. However, instructive parallels can be drawn with the giant planets and Saturn's largest moon, Titan, as well, despite their more extreme contrasts with the terrestrial environment.

Comparing another planet to the Earth is a kind of experiment, almost the only one we have in the physical world where altering the Earth is difficult and inadvisable (one such inadvertent experiment over Antarctica, resulting in the ozone hole, has already been discussed in Chapter 8). All the major bodies of the solar system except Pluto have now been visited and we know in general terms what they are like. This is not the same as saying we understand how they behave, much less that we understand why they behave the way they do. For example, why is Venus so hot and so dry? Are its thick, sulphurous cloud layers created and maintained by massive amounts of active volcanism? How was Mars able to have free water on its surface in the past, when this would be impossible under the conditions we observe today? What mechanism maintains the Great Red Spot on Jupiter, and what is its relationship to the Earth-sized dark spot on Neptune? To what extent do these other-worldly storms resemble terrestrial hurricanes? Simple models and calculations, or attempts to draw analogies with the behaviour of the terrestrial ecosystem, do not readily give definite answers to questions like these. Until we have developed a more complex and complete understanding of climate systems in general, we cannot have confidence in predictions of the behaviour of the climate on the Earth.

The greenhouse effect is a good example of an important phenomenon that is not unique to the Earth. It plays a major role in determining the surface environment on the other terrestrial planets with atmospheres, Mars and Venus, and many of the same greenhouse gases (principally carbon dioxide and water vapour) are responsible. These planets offer an important opportunity to see how the greenhouse effect works in situations other than that we observe on the Earth. The message they send us, that Earth-like planets apparently can be subject to extremely large and variable greenhouse warmings, is an important and currently under-rated one.

## 12.2 Mercury

The atmosphere on Mercury is too thin to have a significant effect on the climate. Instead, the surface temperature is dominated by the interaction of the solar flux directly with the surface. Some interesting complexity is introduced by the large eccentricity of the orbit, and the unusual synchronisation of the rotation of the planet with its orbit,[126] such that a day, as measured by the sunrise, lasts 2 yrs (Table 12.1).

If we simplify the real situation and assume that the surface of Mercury is smooth, and has the same albedo (reflectance integrated over wavelength) everywhere, then it is straightforward to perform an energy balance calculation similar to that for the Earth in Chapter 7 and to derive the variation of temperature with time, as shown in Fig. 12.1. The complexity of the curves, despite the simple 'billiard ball' model of the planet, with no topography and the same surface properties everywhere, arises because of the spin-orbit resonance and the highly eccentric orbit. Note, for instance, that the Sun rises again briefly after setting at some longitudes.

In reality, variations in composition, albedo, and topography will mean that the actual temperature distribution and its variations are even more complicated. The most remarkable example of this is found at the poles, where radar imaging by Earth-based observatories suggests deposits of ice, more than 50 m thick, in craters that are permanently shaded from the Sun. The origin of these is a mystery, since, even without direct exposure to the

[126] Mercury rotates three times for every two orbits around the Sun in an elliptical orbit with an eccentricity of 0.21.

**Table 12.1** Physical parameters for the terrestrial planets

|  | Mercury | Venus | Earth | Mars |
|---|---|---|---|---|
| *Orbital and rotational data* | | | | |
| Mean distance from Sun (km) | $5.79 \times 10^7$ | $1.082 \times 10^8$ | $1.496 \times 10^8$ | $2.279 \times 10^8$ |
| Eccentricity of orbit | 0.2056 | 0.0068 | 0.0167 | 0.0934 |
| Obliquity (°) | 0 | 177 | 23.45 | 23.98 |
| Orbital period (Earth days) | 87.97 | 224.701 | 365.256 | 686.980 |
| Rotational period (h) | 1,407.5 | 5,832.24 | 23.9345 | 24.6229 |
| Solar Day (Earth days) | 175.88 | 117 | 1 | 1.0287 |
| Solar Constant (kW m$^{-2}$) | 10.4[127] | 2.63 | 1.37 | 0.593 |
| *Planet* | | | | |
| Albedo | 0.12 | 0.76 | 0.3 | 0.25 |
| Mass (kg) | $3.302 \times 10^{23}$ | $4.870 \times 10^{24}$ | $5.976 \times 10^{24}$ | $6.421 \times 10^{23}$ |
| Radius (km) | 2,439 | 6,051.5 | 6,378 to 6,357 | 3,398 |
| Surface Gravity (m s$^{-2}$) | 3.70 | 8.60 | 9.78 | 3.72 |
| *Atmosphere* | | | | |
| Principal Composition | He, Na | $CO_2$ | $N_2, O_2$ | $CO_2$ |
| Mean molecular weight (amu) | — | 43.44 | 28.98(dry) | 43.49 |
| Mean surface temperature (K) | $\sim$400 | 730 | 288 | 220 |
| Mean surface pressure (bar) | $<10^{-12}$ | 92 | 1 | 0.007 |
| Mass (kg) | $\sim10^4$ | $4.77 \times 10^{20}$ | $5.30 \times 10^{18}$ | $\sim10^{16}$ |

[127] In the case of Mercury this is the mean of a very large range: the insolation varies between 6.3 and 14.5 kW m$^{-2}$ around the orbit.

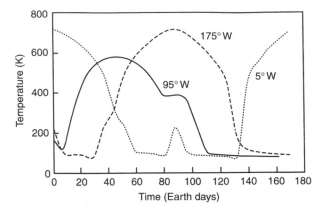

**Fig. 12.1**
The variation during the solar day of the
temperature at the equator on the surface
of a simple, homogeneous model of
Mercury. The three curves correspond to
three different longitudes. Some locations
around the planet get hotter than others
because of the very eccentric orbit and 3 : 2
spin-orbit coupling, and at some locations
an observer would see the Sun rise and set
twice during a full diurnal cycle. (ESA.)

Sun, the ice should sublime away much faster than it can be replenished by
any known source. The prospect of future vacationers skiing on Mercury
has already been note by speculative writers.

## 12.3 Venus

Venus has a particularly interesting climate, because it is Earth-like, but with
an extreme case of the greenhouse effect. When the first spacecraft missions
in the 1960s confirmed findings by radio astronomers that temperatures on
Venus are higher than the melting points of lead, zinc, and tin, it posed a
major challenge to the theoreticians to explain how these are produced and
maintained, one that has not been fully answered to this day. The mean
surface temperature on Venus is around 730K, in spite of the fact that the
net solar input is significantly less than for the Earth, because the very high
albedo of Venus (0.76 compared to around 0.30 for Earth) more than offsets
its greater proximity to the Sun (Table 12.1).

It is still not completely clear how Venus' atmosphere traps the incoming
sunlight so efficiently, but the clouds that cover the planet seem to play a
major role. Not only are there more clouds on Venus than Earth, but the
composition is different, being (in the upper layers at least) mostly concentrated sulphuric acid on Venus. Sulphuric acid cloud droplets scatter
sunlight very conservatively, that is, with little absorption, diffusing a fraction of the incoming sunlight down to the surface. At the same time, they are
also very opaque in the infrared, so clouds made of concentrated sulphuric
acid droplets have a higher optical depth in the thermal infrared than they
do in the near infrared and visible part of the spectrum. The cloud blanket
therefore makes a large contribution to the backwarming effect, which outweighs the albedo effect overall when combined with the contribution of
around a million times as much $CO_2$ as the Earth, plus substantial traces of
water vapour and other greenhouse gases.

As we did for the Earth, we can investigate the Venusian climate with
simple greenhouse models, starting with the single-layer atmospheric model
described for the Earth in Section 7.5 and Fig. 7.5. The solar constant $S$ at
Venus is about twice that at the Earth, at 2,626 W m$^{-2}$. Assuming an albedo
of $A = 0.76$ and dividing by four, the ratio of the area of the spherical planet

to its cross-sectional area (Fig. 7.3), the mean solar power absorbed by unit are of the atmosphere is therefore 157 W m$^{-2}$. Equating this to the outgoing blackbody flux $\sigma T_e^4$ (where $\sigma =$ Stefan's constant and $T_e$ is the effective radiating temperature of the planet, cf. Section 7.4) we obtain $T_e = 230$K. As we would now expect, this is much lower than the surface temperature of Venus, which is 730K, and is in fact approximately the temperature at the cloud tops.

In the single-slab atmospheric model, the mean flux downwards at the surface is

$$F_\downarrow = \frac{S(1-A)}{4} + \sigma T_e^4 = 2\sigma T_e^4 = 314\,\text{W m}^{-2},$$

where we have again assumed that all of the solar energy not reflected back into space is absorbed at the surface, that is, neglecting atmospheric absorption at visible and near-infrared wavelengths. For balance the downward flux at the bottom of the atmosphere must be the same as the upward flux from the surface at temperature $T_0$, so

$$T_0 = \left(\frac{314}{\sigma}\right)^{1/4}$$

from which $T_0 \sim 273$K. This is less than half of the observed surface temperature of Venus, showing that the simple one layer model is, not surprisingly, quite inadequate for treating a thick atmosphere with a large vertical temperature gradient.

The more sophisticated model depicted in Fig. 12.2 represents the atmosphere as a stack of $N$ optically thick, isothermal layers. If we continue

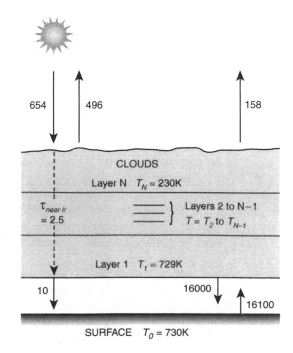

**Fig. 12.2**
An energy balance model for Venus, showing radiative fluxes in W m$^{-2}$. The main cloud layers have a high reflectivity in the near infrared and visible parts of the spectrum, and a high absorptivity in the thermal infrared. For the amount of energy escaping as thermal radiation from below the clouds to match that absorbed from the Sun, the temperature at the surface has to become very high.

to treat the cloudy atmosphere as reflecting but non-absorbing at solar wavelengths, the energy balance at the surface is now given by

$$\sigma T_0^4 = \frac{S(1-A)}{4} + \sigma T_1^4,$$

which gives the correct answer $T_0 = 730K$ if $T_1$, the temperature of the lowest atmospheric layer, is just 2K cooler than the surface, that is, $T_1 = 728$ K. The first term on the RHS, the direct solar input, is only about 1% of the downward flux from the atmosphere. (In fact, measurements by radiometers on atmospheric probes on Venus have shown that the real number is even smaller, about 0.1%, since absorption of solar radiation by the clouds and gases in the atmosphere is not negligible, but actually corresponds to an optical depth of about 2.5. However, this makes only a small difference to the value we deduce for $T_1$.)

For the topmost layer of the troposphere, neglecting the opacity of the stratosphere, which is small at most wavelengths, we have for balance

$$\frac{S(1-A)}{4} = \sigma T_N^4,$$

whence $T_N$ is equal to the effective emitting temperature $T_e$, with a value of 230K. The other layers, 2 to $N-1$, have progressively reducing temperatures from somewhat less than 730 to somewhat higher than 230K. Since all layers are optically thick, the temperatures are determined by convective equilibrium, with $dT/dz = -g/c_p$, which for the value of $c_p$ corresponding to $CO_2$ gives a lapse rate of about $10K\,km^{-1}$. This specifies the temperature at every height, given that of the surface, and tells us that the vertical distance between the surface at 730K and the cloud tops at 230K is approximately $(730-230)/10 = 50\,km$, which is quite close to the observed height of around 60 km (cf. Fig. 3.9).

Above the cloud tops, the atmosphere is optically thin in the thermal as well as the solar part of the spectrum and we expect an isothermal profile, at a temperature $T_{strat}$ given by (cf. Section 3.5.2)

$$T_{strat} = \frac{T_e}{2^{1/4}} \approx 193K,$$

which again is not far from that observed.

This sort of calculation shows that, as with the Earth, we can account for the observed thermal structure of Venus' atmosphere by a combination of thermal balance at the surface, with space, and in the stratosphere, and convective equilibrium in the troposphere. The high surface temperature results from the very deep and thermally opaque atmosphere containing large quantities of efficient greenhouse absorber, in particular $CO_2$ and $H_2SO_4$.

It still remains to explain the exceptionally high opacity and its spectral variation in detail, and of course why the Venusian atmosphere is so massive in the first place. It has been calculated that the amount of $CO_2$ that once was in the Earth's atmosphere and is now locked in carbonate deposits after having first dissolved in the ocean, is of the same order as that which is still in the atmosphere on Venus. The implication is that the Earth, by being cool enough to condense liquid water on its surface, escaped having a much thicker $CO_2$ atmosphere. Also, Venus may have been much more volcanically active than Earth, and even today volcanoes may be pumping large amounts of carbonic and sulphurous gases into the atmosphere, continually 'topping up' the huge greenhouse effect. This could, in time, subside.

## 12.4  Mars

The atmosphere of Venus is $\sim$100 times thicker than Earth's; the atmosphere of Mars is $\sim$100 times thinner. The gaseous component of the Martian atmosphere is, like Venus, mainly $CO_2$, but with a mean surface pressure of only about 0.06% of the Earth's. Even this relatively small amount of gas, when combined with the effect of airborne dust, is enough to produce a measurable greenhouse effect, which again we can estimate with a model. Taking the values for solar constant and albedo from Table 12.1 to be $S = 593\,\mathrm{W\,m^{-2}}$ and $A = 0.25$, we find

$$T_e = \left( \frac{S(1-A)}{4\sigma} \right)^{1/4} = \left( \frac{593(1-0.25)}{4 \times 5.67 \times 10^{-8}} \right)^{1/4} = 210\mathrm{K}$$

for the effective radiating temperature of Mars. Compared to an observed mean surface temperature of 220K, this indicates a small greenhouse effect of about 10K. The predicted stratospheric temperature is

$$T_{\mathrm{strat}} = \frac{T_e}{2^{1/4}} = 177\mathrm{K}$$

and, since the adiabatic lapse rate $\Gamma = g/c_p$ is 4.5K $\mathrm{km}^{-1}$ on Mars, the tropopause should fall at approximately

$$z_{\mathrm{trop}} = \frac{(250 - 177)}{4.5} = 16.2\,\mathrm{km}$$

above the surface. The profiles constructed from these numbers are quite close to the left-hand pair of temperature profiles in Fig. 12.3, which were computed using a more detailed radiative-convective model for Mars, assuming a pure $CO_2$ atmosphere.

However, the first detailed survey of Martian atmospheric temperatures, by the first artificial satellite of Mars (Mariner 9, in 1971), produced results that were rather different from these expectations. The measured profiles all fell inside the shaded area in Fig. 12.3, and so were much warmer than expected and had a much shallower lapse rate. The reason is that the main contributor to the Martian greenhouse effect is not carbon dioxide but airborne particles, in the form of windblown dust from the surface. The effect of dust is to raise surface and air temperatures several tens of degrees relative to clear conditions, as shown by the second set of calculated profiles in Fig. 12.3, which include a model for the dust opacity. Under very dusty conditions, the greenhouse effect can increase five or even ten-fold over

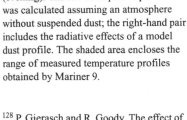

**Fig. 12.3**
Calculated temperature profiles for the atmosphere of Mars,[128] at two local times of day, 0600 (morning) and 1800 (evening). The left-hand pair of profiles was calculated assuming an atmosphere without suspended dust; the right-hand pair includes the radiative effects of a model dust profile. The shaded area encloses the range of measured temperature profiles obtained by Mariner 9.

[128] P. Gierasch and R. Goody. The effect of dust on the temperature of the Mars atmosphere, *Journal of the Atmosphere Sciences*, **29**, 400–402, 1972.

the clear situation, raising surface temperatures above the freezing point of water at temperature latitudes.

Under dusty conditions, when the infrared opacity is high, we might expect the single-slab atmospheric model of Section 7.5 to give a reasonable approximation to the greeenhouse effect. According to this model, in which the atmosphere is optically thin in the visible and optically thick in the infrared, we obtain

$$T_0 = \left(\frac{2S(1-A)}{4\sigma}\right)^{1/4} = 250\text{K}$$

for the surface temperature $T_0$, and again this is close to the more refined model prediction, and is also a reasonable match to the average value of the Mariner 9 observations.

Two other things worthy of note that stand out in Fig. 12.3 are the large diurnal temperature changes near the surface, and the non-adiabatic lapse rate in the dusty troposphere. The former arises because of the low density of the Martian atmosphere, even at the surface, and its composition of almost pure $CO_2$. In the absence of dust, the atmospheric opacity in the infrared is dominated by the few strong bands of $CO_2$ (especially that at 15 $\mu$m), which are very opaque. Since there is very little cloud, water vapour, methane, or ozone to plug the gaps between these bands, the spectral windows on Mars are wide and clear (the opposite of Venus where the high densities and rich compositional mix leaves very few gaps). As a result, the surface not only receives most of the incoming solar flux directly, but also is able to cool efficiently to space via the windows in the infrared spectrum. The atmosphere, by contrast, is optically thick in the radiating $CO_2$ bands until stratospheric altitudes are reached and has to cool mainly by convection. This gives rise to the usual adiabatic lapse rate in the troposphere when dust is absent. The explanation for the sub-adiabatic lapse rate in the dusty case is simply that the solar energy is then no longer deposited mainly at the surface, since the dust particles are good absorbers in the near infrared (again in contrast to the conservatively scattering $H_2SO_4$ cloud droplets on Venus), as well as at the longer thermal infrared wavelengths. Thus, there is heating in the dusty Martian tropopause at all heights, and radiative-convective, rather than just purely radiative, equilibrium prevails.

In situations where the dust loading of the air varies from place to place, the resulting horizontal temperature contrasts produce pressure gradients that drive strong winds. These raise more dust and so produce positive feedback, sometimes leading to enormous planet-wide dust storms that rage for months. The Martian climate is therefore a dynamic affair, with big seasonal changes in the $CO_2$ part of the greenhouse, and a global dust contribution that varies even more and on time scales of as little as a day.

Mars also possesses the solar system's most dramatic evidence of past climate change. Today's frozen rocks and desert bear unmistakable features of rivers, lakes, and seas. The origin of these is still controversial, but one leading theory is that Mars may in the past have had a much thicker atmosphere. This would still have consisted primarily of $CO_2$, but being warmer would have held much more water vapour, and possibly contained other constituents including substantial clouds. Together these might have produced a greenhouse effect large enough to raise the temperature and pressure to values that could support liquid water on the surface.

If so, how could Mars have changed so much since then? The answer may lie in the high eccentricity of its orbit and the large fluctuations in insolation that result from resonances between this and other orbital parameters (the Martian version of the Milankovich cycles, Section 10.2). Alternatively, the heating could have been primarily due to volcanic activity, which has since subsided. In this case, water vapour, $CO_2$, and other gases and particles from the interior of Mars could have warmed the planet by greenhouse action and produced lakes and rivers for as long as the volcanoes were sufficiently active.

Yet another possibility is that Mars lost most of its early atmosphere in collisions with other large bodies, such as those we see in more stable orbits in the asteroid belt today. The reduced greenhouse warming could have seen temperatures fall below the freezing point of water long and often enough to remove water vapour and clouds from the remaining atmosphere. This would further reduce the surface temperature and eventually lead to the situation we find today, where the water not lost in the collision mostly lies frozen beneath the surface and in the polar caps.

## 12.5  Titan

Saturn's giant moon Titan is larger than Mercury, and not much smaller than Mars, but boasts a thicker atmosphere even than that of the Earth with a surface pressure of around 1.5 bars, and with the same primary constituent, nitrogen. Titan is ten times more distant from the Sun than Earth and so very cold. It has no free oxygen—nobody expects to find life—but a lot of methane and other hydrocarbons, some of which form oily haze layers composed of light hydrocarbons. A steady drizzle from these, plus the probability of rain from the lower-level methane clouds, mean that an ocean of light oil on the surface is a real possibility (Fig. 12.4).

At Saturn's mean distance from the Sun, the solar constant is only about 1% of that at Earth. Titan has an albedo of 0.2, which works out, using the

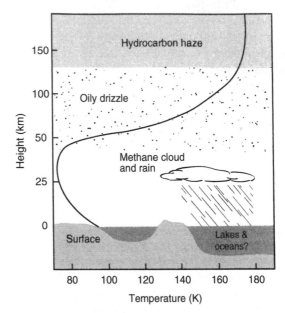

**Fig. 12.4**

The main features of the climate system on Titan, as they were understood prior to the Cassini-Huygens mission, including an atmospheric temperature profile inferred from the occultation of the Voyager 2 probe as it flew behind Titan in 1980.

simple energy balance model (Section 2.9), to an effective emitting temperature of 85K. Again applying the one-layer greenhouse model (Section 7.5), we have

$$T_0 = \left(\frac{2S(1-A)}{4\sigma}\right)^{1/4} = 99\text{K}$$

for the surface temperature $T_0$, and

$$T_{\text{strat}} = \frac{T_e}{2^{1/4}} = 70\text{K}$$

for the stratospheric temperature $T_{\text{strat}}$. If the dry adiabatic lapse rate $\Gamma(g/c_p = 1.2\text{K km}^{-1}$ on Titan) applies, then the tropopause should fall at approximately

$$z_{\text{trop}} = \frac{(99-70)}{1.2} = 24\,\text{km}$$

above the surface. These global mean values compare reasonably well to the profile in Fig. 12.4, which is based on the radio occultation profile measured near the equator on Titan by Voyager 2 in 1980.

## 12.6 The Jovian planets

Climate, in the sense in which we understand it on the terrestrial planets, is not really defined on the gas giants of the outer solar system, which lack solid surfaces, except possibly at great depths where the pressure and temperatures are so high as to take us into completely unfamiliar physical regimes where, for example, even hydrogen is stripped of its bound electrons and has to be reclassified as a metal.

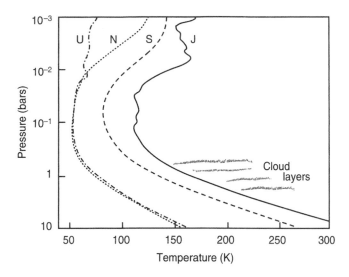

**Fig. 12.5**
Temperature profiles for the four giant
outer planets, measured by the NASA
Voyager spacecraft using infrared
sounding and the radio occultation method.

Nevertheless, the outer regions of the atmospheres of these large planets
do have almost Earth-like temperature and pressure regimes (Fig. 12.5).
These are produced by processes related to those that are familiar to us
on our planet, for instance, the absorption and transmission of sunlight in
gaseous absorbers and clouds, and the transfer of latent heat as condensates.
Along with quite different atmospheric compositions (mainly hydrogen and
helium, with substantial traces of ammonia, methane, and water) the main
difference from the Earth-like inner planets is that they have internal sources
of energy that are of the same order as that arriving externally from the Sun.

Taking Jupiter as the largest, closest and therefore best-studied example,
we calculate from its distance from the Sun (five times that of Earth) and
albedo (0.35) that its effective emitting temperature should be

$$T_e = \left(\frac{S(1-A)}{4\sigma}\right)^{1/4} = \left(\frac{50.5(1-0.35)}{4 \times 5.67 \times 10^{-8}}\right)^{1/4} = 110\text{K}$$

if it is in balance with the Sun. If we add the internal flux of about
$30\,\text{W m}^{-2}$, then

$$T_e = \left(\frac{S(1-A)}{4\sigma}\right)^{1/4} = \left(\frac{50.5(1-0.35)+30}{4 \times 5.67 \times 10^{-8}}\right)^{1/4} = 130\text{K},$$

which is close to the observed value. The predicted stratospheric temperature is

$$T_{\text{strat}} = \frac{T_e}{2^{1/4}} = 109\text{K},$$

which is also close to the measured value (Fig. 12.5). The heat flux from the
interior, which is produced by the conversion of potential energy as the gas
giant contracts slowly and heavier elements move towards the core, along
with the high opacity of the atmosphere in the thermal infrared at pressures

higher than about 1 bar, guarantees that the deep atmosphere is convective, with a lapse rate close to

$$\Gamma = \frac{g}{c_{\mathrm{p}}} = \frac{0.02425\,\mathrm{km\,s^{-2}}}{0.01236\,\mathrm{J\,kg^{-1}\,K^{-1}}} \approx 2\mathrm{K\,km^{-1}}.$$

## 12.7  Planets of other stars

Because our own solar system exists, and given the size of the Universe with an estimated 50 billion galaxies, some containing thousands of billions of stars, the probability is that planetary systems exist in large numbers.[129] Some of these have already been detected, more than 100 at the time of writing, although with a bias towards those that contain very large planets—more like Jupiter than Earth—because these are much easier to detect.

It will be only a matter of a decade or two until we are able to detect planets similar to the Earth, orbiting stars that resemble the Sun in size and luminosity. Very large telescopes, floating in space, have already been designed that have the capability to obtain infrared spectra of these planets, allowing techniques like those described in Chapter 9 to be employed to infer the existence of atmospheres, and their compositions and temperatures. Assuming reasonable technical progress, and the expenditure of sufficient funds, basic details of the climate on a number of these new Earths may be available to us in a few decades from now. It is likely that these will be sufficiently detailed to reveal an Earth-like environment where these exist.

Figure 12.6 shows what Venus, Earth, and Mars would look like when viewed from outside the Solar System with a spectrometer attached to a suitably large telescope. The dominant feature in all three is the carbon dioxide band at 15 μm, while the 6.3 and 9.8 μm water and ozone bands

[129] 'With $10^{11}$ stars in our galaxy and $10^9$ other galaxies, there are at least $10^{20}$ stars in the universe. Most of them may be accompanied by solar systems. If there are $10^{20}$ solar systems in the universe, and the universe is $10^{10}$ years old—and if, further, solar systems have formed roughly uniformly in time—then one solar system is formed every $10^{-10}$ yr = 3 ms. On the average, a million solar systems are formed in the universe each hour'. (Carl Sagan.)

**Fig. 12.6**
Low-resolution infrared spectra of Venus, Earth, and Mars, based on actual observations from spectrometers on NASA spacecraft. In the next few decades, we should have spectra of this quality from planets orbiting other stars, and from them we should be able to say whether ozone (hence oxygen) and water are present, and possibly to estimate the surface pressures and temperatures.

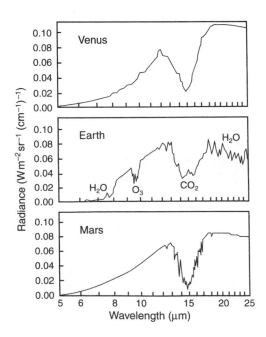

only show up strongly on the Earth. Such a telescope—actually an array of telescopes, floating in space and co-located using laser inteferometry—called DARWIN[130] is currently planned by the European Space Agency. An 800-hour (33 days) exposure is required to obtain a spectral resolution of 0.5 μm per pixel with just enough signal-to-noise ratio to permit the detection of water and ozone with reasonable certainty. The presence of large amounts of ozone suggests that oxygen is also present and would be a strong indication of the presence of active life on the planet.

## 12.8  Planets without stars

There are likely to be some—possibly a vast number, but we cannot detect them at present—planets that do not orbit stars at all, but drift in the darkness of space. If one of these was identical to the Earth, but had a thick hydrogen atmosphere (again a realistic possibility, since our Earth probably once had such an atmosphere, that was stripped away by the Sun) then a curious situation is possible. The Earth emits a small amount of internal heat,[131] due to cooling and radioactive decay. Hydrogen at high pressure is very opaque at all infrared wavelengths, and so provides greenhouse-like trapping of this internal heat. Only around 50 bars of surface pressure are enough to raise the surface temperature to around 300K, so this drifting Earth, despite permanent darkness and less than ideal atmospheric conditions from a human perspective, could enjoy liquid water oceans on its surface.

### Further reading

*The Physics and Chemistry of the Solar System*, J.S. Lewis, Academic Press, New York, 1995.

*Theory of Planetary Atmospheres*, J.W. Chamberlain and D.M. Hunten, 2nd edition, Academic Press, New York, 1987.

*Planetary Sciences*, Imke de Pater and Jack J. Lissauer, Cambridge University Press, Cambridge, 2001.

[130] Darwin is a flotilla of six 1.5 m telescopes floating in space about 100 m apart, with a node spacecraft that combines the beams to make a large optical interferometer, and a housekeeping spacecraft that relays the data, all orbiting the Sun at a Langrangian point, which keeps it at a fixed distance from the Earth. A similar project called *Terrestrial Planet Finder* is under development in the USA.

[131] Actually, about 30 trillion watts.

### Questions

|                                              | Venus | Earth | Mars |
|----------------------------------------------|-------|-------|------|
| Distance from the Sun ($10^9$ m)             | 108   | 150   | 228  |
| Bolometric albedo                            | 0.76  | 0.3   | 0.25 |
| Acceleration due to gravity (m s$^{-2}$)     | 8.88  | 9.81  | 3.73 |
| Specific heat at constant pressure (J kg$^{-1}$ K$^{-1}$) | 830   | 1005  | 830  |
| Surface temperature (K)                      | 730   | 290   | 250  |

1. Using the data in the table and an effective temperature of 6,000K for the Sun, which has a radius of 700 Mm, calculate the effective temperatures and the tropospheric lapse rates of Venus, Earth and Mars. Sketch the vertical temperature profiles for the three planets, showing explicitly your calculated values for the stratospheric temperatures and tropopause heights.

2. Estimate the surface temperature on Venus if the volcanic activity ceases and the cloud cover vanishes as a result. Repeat the calculation allowing for the estimated effect of removing the cloud on the backscattering of solar radiation to space. Assume that the mean optical depth of the cloudless atmosphere in the thermal infrared is about 10, and that the albedo $A$ of non-cloudy Venus is likely to be more like that of Mars, say $A = 0.2$.

3. The distance of Jupiter from the Sun is $780 \times 10^9$ m, and its bolometric albedo is 0.35. Calculate an effective emitting temperature for Jupiter, and explain why this differs from the measured value of 125K. What physical processes inside Jupiter might be responsible?

4. The largest satellites of the giant planets Jupiter, Saturn, and Neptune—Ganymede, Titan, and Triton, respectively—are comparable in size to the terrestrial planets, but only Titan has a substantial atmosphere. Discuss some possible reasons for this.

# Epilogue

Everyone knows that climate is a topic that is attracting a great deal of attention, among powerful politicians and the general public as well as scientists. This is hardly surprising, given the stakes involved if the climate change that occurs in the next 100 yrs is as bad as the current best predictions say. Misunderstandings are rife, however, along with controversy and rash predictions, of both the doom-laden and the overly complacent variety. The student of climate is expected to shed light on the issues and debates that will help progress towards a productive consensus.

A number of Frequently Asked Questions are important 'drivers' for climate physics research. Perhaps top of the list of FAQs are: (a) whether human activities are responsible for a significant amount of climate change at the present time, (b) what will be the impact of the changes if they continue, and (c) what must we do to ameliorate them?

Even simple models indicate that the changes we have seen in the composition of the global atmosphere since the industrial revolution are more than enough to produce the changes in mean surface temperature that we observe. The most complex models show a very good fit between anthropogenic forcing and the actual behaviour of the climate. Of course, the models may be wrong; they almost certainly are since we are aware of numerous inadequacies in their formulation and execution, but of course the effect of these is to increase the error bars and not the sense of their predictions.

Those who remain sceptical generally fall into two major categories. In the first are those who think the problem is just too complex and difficult—after all, we are trying to predict the fairly distant future—and who note the high cost of limiting emissions or population growth in an attempt, that will probably be nugatory, to address a problem that may not come to pass anyway. Opponents of this philosophy call it the 'head in the sand' approach. In the second category are those who know some science; they note (correctly) that there are large natural fluctuations in the climate, but then jump from that knowledge to the conclusion that there is therefore no risk of anthropogenically produced changes in addition, when the opposite would be more sensible. Others discover the existence of feedbacks—for example, the negative feedback loop corresponding to warmer oceans → more water vapour → more clouds → higher albedo → cooler surface. While again it is clearly important (and difficult) to account for feedback, both positive and negative, the assumption that feedback effects will exactly, or nearly, cancel out any new forcing is again illogical.

These views, and other combinations and variations, are widely and sometimes fervently held, but the majority, among scientists at least, is now convinced that human-induced global warming of a serious kind is more likely than not in the next 50–100 yrs. What constitutes a 'serious' impact of climate change?

The question of impacts is a huge one—the IPCC has one sub-commission on the science of global warming (the one whose results we have been quoting) and another, separate one on impacts (and a third studying amelioration, the politics of responding to the threat)—and we cannot possible do justice to it here. It is, in any case, a subject which, in universities, is usually found in geography not physics departments. Briefly, however, we have seen that sea level rises induced by climate change are likely to be in the range from 0.1 to 1 m. The impact obviously is greatest in low-lying coastal regions, and it is estimated that 100 million people worldwide live within 1 m of sea level.

Current climate models all predict increased average precipitation accompanying increases in temperature. However, the increase is concentrated in regions that already have heavy rainfall; in rain-starved regions the rainfall is predicted to get lighter, as a result of strengthened circulation cells, which extended to higher, cooler regions in the atmosphere where the humidity is low even at saturation. The implication is that floods and droughts will be even more of a concern than at present.

A warmer, wetter world will have an impact on the ocean circulation, and at some stage higher rainfall will introduce more fresh water and reduce the salinity, particularly at higher latitudes. The predicted effect in coupled atmosphere–ocean models is a weakened circulation, possibly eventually leading to a rapid change between equilibrium states of the kind that appear to have occurred in the past.

Some beneficial impacts are also possible; longer growing seasons at higher latitudes along with the stimulus that increased carbon dioxide provides may lead to increased global food yields, for example.

Despite an inevitable focus at the present time on potentially harmful global change, climate physics is not a gloomy subject. By understanding the climate system better, and learning how to predict its response to forcing, we can move towards a solution to greenhouse-induced global warming. No doubt this will involve clean technology, perhaps an advanced version of the present primitive nuclear power industry, possibly based on fusion rather than fission, and made widely and cheaply available, so wasteful combustion-based technologies will be left behind.

Since the passage of the Montreal Protocol limiting the emission of CFCs into the atmosphere, ozone depletion is showing signs of slowing down and recovering. The current level of global depletion, now around 4 or 5%, might have been twice as large without the restrictions on chlorine emissions imposed by the protocol. Complete recovery of the ozone layer has been forecast by 2050 if the present rate of progress is maintained; UN Secretary-General Kofi Annan praised the Montreal Protocol as 'perhaps the most successful environmental agreement to date'.

Working against better technology and international cooperation is the sharply increasing demand for energy. Of the total carbon dioxide currently generated by burning fossil fuels, about 20% comes from transportation and the rest from static sources, like houses, factories, and power stations.

The human population of the world is increasing by about 100 million every year, and the average standard of living, and so the per capita energy consumption, is also increasing. What prospect is there for meeting this need? No one knows, but here is a clue: we saw in Chapter 2 that that the Sun emits enough energy in one second to supply the Earth's energy needs for nearly a million years.

# Index